DEAD EPIDEMIOLOGISTS

On the Origins of COVID-19

DEAD EPIDEMIOLOGISTS

Rob Wallace

MONTHLY REVIEW PRESS

New York

Library of Congress Cataloging-in-Publication data
available from the publisher

ISBN paper: 978-158367-902-9
ISBN cloth: 978-158367-903-6
ISBN eBook: 978-158367-904-3

ON THE COVER: Jennifer McQuiston, Jonathan Towner and Brian Amman
approach a bat cave in Queen Elizabeth National Park (Uganda) on August 25,
2018. Amman and Towner, CDC scientists, gathered twenty bats that reside in
Maramagambo Forest as part of a research project to determine flight patterns
and how they transmit Marburg virus to humans. Bonnie Jo Mount/*The
Washington Post* via Getty Images.

Typeset in Minion Pro and Brown

MONTHLY REVIEW PRESS, NEW YORK
monthlyreview.org

5 4 3 2 1

Contents

Chen could see bits and pieces of the future, "but only in equations." A frequent lament. Numbers could attack the flesh, the will, but rarely built it up. Morale for them never lay in the numbers. He made poetry out of his premonitions, his equations, because they'd proven useless to him as fact, because he was never sure whether he was actually seeing the past.

—Jeff VanderMeer (2019)

Preface

When I tell you I'm Mexican, I'm pointing to the country that wanted me dead first. When I tell you I feel American, I'm telling you I often think about how I'll die here. Sometimes, when we cite our geographies, we're telling you what we survived and what is going to kill us.

—Miss Jesús (2020)

LATE MARCH THIS YEAR, I'm lying on my bed gasping for breath, trying to catch up on the day's deoxygenation, a plane slowly falling out of the sky. A half hour into the evening's panting—and nothing so pleasurable as the noise implies—it's a bit of a wonder that I am now what I study. I've got COVID-19 and in my lungs something whose ancestor circulated among bats in greater Yunnan, on the other side of the planet, maybe only the year before. Millions sheltering in place, our worlds are both smaller than at any other time in most of our lives and, with this specter from half a world away haunting infected and well alike, too large from which to hunker down. We're more online than ever, seeing almost nothing of our neighbors, the busiest streets ghost towns to the end of the empty light rail line. Through May, before the revolt and demonstrations here in the States, our social geometries were folded inside out, with all the trauma of a snake whose live meal ate its way out into the clear.

Where did I pick up my infection?

Earlier in March, I traveled to a premonitory of a conference on racial injustice and health held in Jackson, Mississippi. I flew in to New Orleans a few days earlier—cheaper tickets—taking in a quiet night on the town I hadn't visited in twenty-seven years. A career along the poverty line, this middle-aged man stayed at a youth hostel before taking a Greyhound north through a part of the country he hadn't seen before. In the room with two bunk beds lay a European intermittently coughing out his guts. Maybe him? Yeah, a distinct possibility. Or was it in the movie theater a few days earlier in St. Paul when a stye suddenly struck me mid-reel? Or upon my return, at the community event in Minneapolis that a couple soon sick attended, later conscientiously announcing their infections, confirmed by what testing was then—and, ugh, is even now—only kinda available.

COVID-19 in the United States proved canonical. Pandemics slow and fast—HIV and H1N1 (2009)—typically get sucked into New York City early on before being blown back out to the rest of the country by way of the travel network, down a hierarchy of town size, economic power, and travel load.[1] New Orleans was one of the first struck post–New York.[2] I wonder if I did my part, transporting SARS-CoV-2, the COVID virus, to Mississippi, seeding, as documented in other states, what wouldn't be detected by the public health system there for another six weeks.[3]

Our airports and the high-end shopping there have been remodeled as cathedrals to the neoliberal sublime. American bus stations are more packing centers for delivering the poor and working class, making their way—I overheard several conversations in Nawlins and Baton Rouge—to sheltering in place with friends and family against bad partners and bad wages long before the outbreak. One man, just released from jail, with all he possessed in a brown paper bag. Another leaving a marriage for a job. At the other end of something of a spectrum, an oil rigger, chatting up a woman of a certain age, with cash enough to invest in house-flipping. People rough in manner but for the most part fine in spirit, however much any of us manifest our circumstances.

The bus drivers and the station clerks were nearly to a one Black. There were no supervisors on-site (and if could be helped, no cops). The Black working class run actually existing Greyhound on its own, getting various and sundry glitches worked out, including botched schedules and a bus that refused to start through the company smart-phone—high-tech supervision—to which it was slaved. The drivers proved both good-natured and broaching no bullshit. They assured their charges and admonished misbehavior. They steered their buses, machine and people both, to their destinations.

Out of New Orleans, I caught the surprises I planned. I saw cranes roosting on rebar in Gulf wetlands suffusing the concrete pylons atop which the interstate ran. A dilapidated bed-and-breakfast along the biomorphic landscape—"La Belle Maison"—swished by. The electrical grid hopped stanchion to stanchion until it disappeared into the fog of the sea. As if our mode of energy production was indeed infinite.

Jackson, named after the slaver and genocidal general, his statue only now to be removed from in front of City Hall, has been for decades running a center of experimentation in Black liberation. The history and the civil rights museums are stunning tributes to fierce resistance. I learned it was at the station our bus rolled into where twenty-seven Freedom Riders were arrested as soon as they arrived. A few years later an albeit titular Black Republic stretched over multiple counties in goddamn Mississippi. Today one can trace the through line not just for legal rights but self-determination and economic democracy to—in no monolith—Mayors Chokwe Lumumba and Chokwe Antar Lumumba, the Jackson-Kush plan, the Malcolm X Grassroots Movement, the New Afrikan Peoples Organization, June Hardwick, the People's Assembly, Rukia Lumumba, solidarity economies, the Amandia Education Project, the Take Back the Land campaign, Operation Black Belt, participatory budgeting, Tougaloo Community Farm, Adofo Minka, and Kali Akuno and the inimitable Cooperation Jackson.[4]

Our climate and health team broke bread with another local group—One Voice. One of its projects pursues energy democracy.[5] The rural counties of Mississippi are powered by energy cooperatives run by boards of good ol' boys that pass on energy costs to poor Black majorities for as much as half of family incomes, while, among other egregious examples, wiring private schools for free. The One Voice project works out of a building in Jackson—1072 John R. Lynch Street—that was the heart of the civil rights movement. Everybody passed through: Kwame Ture, Ella Baker, Fannie Lou Hamer, and thousands of others. Fifty years later and the fight continues along these new-century axes. One Voice is supporting efforts to break these boards by getting Black people elected. Out beyond a radical Black-run state capital, where the Confederate flag flew at the governor's mansion until only this month, secret meetings and shady deals churn apartheid onward, fought against by an organically organized people's movement. As I offered to a colleague who lives in Jackson on what first appeared an entirely different topic—she was being stymied by a board at a local institution—many white people are terrified of Black excellence.

It isn't all fried okra and barbecue beans. One night in town, I went to an emergency room to get my stye checked out. The bag beneath my eye was filled with fluid, an ugly red. As if someone had decked me. American health care is Second World for most. A Mississippi hospital, run well enough, felt another half-step down. We had to haggle with the ER computer whether my insurance covered out-of-state visits, as if my health stopped at the Minnesotan border. I played a loving prank I've taken up in the past couple years. I prompted the desk nurse, who had an accent that pinged between her Eastern European roots and down-home Mississippi, that this bureaucratic folderol would all go away once we instituted a national health service. Her face brightened at what passes for a paradise of a possibility for American medical staff. Upon filling my prescription, the nurses, kindly to a one, repeatedly warned me to avoid downtown at night.

Days later, back home in St. Paul, I worked hard through my characteristically slow starter of a COVID infection, the first twelve days one long pounding headache and intermittent dizzy spells. I was writing up some of the articles collected in this volume, all finished in the first half of 2020. As my health slowly deteriorated to frightening bouts of shortness of breath and an associated decline in executive function, my kid, at his mother's, yelled at me through the computer to get my shit together and get some help. With the insidious infection bouncing between scary to whatevs, I needed the prompt. Landing one article on the publication tarmac, I unplugged out of my obligations to get my case checked out.

In a classic Minnesota evasion, the administrative assistant at my primary clinic refused to give me a straight answer over the phone whether my doctor would see me. Upon a decade's experience in this kind of rope-a-dope, I figured she was refraining from giving me news she thought I didn't want to hear. And it was certainly that. But the outbreak had broken all my pretense at coddling Jantelagen passiveness. It was my turn to yell. That I needed answers. That not giving them was hurting me more than the disappointment they offered. While I understand why I wouldn't be seen—to keep the clinic from amplifying the outbreak—my doctor, whom I like very much, still owes me an explanation on the other side of the pandemic. Why would I ever trust their judgment again if they're to put the system's needs over mine? A lucrative contradiction beyond us both, tucked into American health care by design.

So I was to be diagnosed online. The first question asked me whether my illness was serious. I clicked "yes" and was bounced to a list of clinics (where I wasn't allowed). It took me several times to realize I had to choose "no," my illness wasn't serious, to reach the COVID landing page. Upon cycling through a series of questions and waiting a day, a nurse practitioner, seeing neither hide nor hair (nor nasal sample), confirmed a diagnosis. No test. No antiviral. No masks and no gloves provided. No community health practitioner stopping by to check on me. Abandoned by modern medicine and the State. Of course, my initial anger was misdirected at the scheduler, born of fear and frustration of a frayed empire that felt beyond intervention. Was it any wonder I was thrilled when the corrupt Third Precinct was burned and looted across the river in Minneapolis two months later? Someone—a multiracial subaltern subjected to the worst of it—had hurt the system back.

I did remember I had an N95 mask somewhere in the back of the utility closet. I've been studying infectious diseases since my early days in graduate school. I'd been working on the new generation of threats since 2005, starting on bird flu H5N1. Over this expanse, I've had to learn to compartmentalize myself from the dark logic that pulses at the heart of capitalist land use and agricultural development, selecting, along its aggressively expanding gyre, for the most danger-ous pathogens possible. Every once in a while, however, I'm shook by the reality behind the numbers and narratives dancing across my screen. The emergence of avian influenza H7N9 disturbed me enough to purchase the mask. So out-of-step Rob circa 2013 helped sourly vindicated Rob 2020. Just the kind of precaution the world govern-ments, servicing first and foremost the very industries driving the emergence of these pathogens, offered only a desultory hand signal. Did it mean "a-okay" or "white power" or both? Dead epidemiolo-gists played doctor the past decade, jet-setting about in that gesture's negative space, rewarded for administering what, in actuality, was a grand fiasco. Not a-okay.

The collection here addresses the resulting COVID-19 outbreak in near-real time. Keep track of the dates of publication across a dilated duration, within which each week still seems a month, with new rev-elations about COVID-19 emerging on the daily. It's surprising what twenty-five years of study can do to help prep one for such an out-break. I had wired most of what was going on by mid-January. The first essay here reads as a veritable projection of what would indeed accrue the world over in the months to follow. At the same time, we can detect a shift in understanding across the pieces beyond assimi-lating new data. A broader conceptual framework now fits the strange menagerie of pathogens that have emerged this century, one after another.

The origins of the COVID outbreak are tracked over and again to their multiple moments of inception, starting in December 2019, yes, but also in and out of multiple biocultural domains, to capitalism and science more broadly and, without a whit of prepper snark, to the very beginnings of civilization. We end the collection in the bat caves of COVID's proximate start, where many might think we'd begin instead. But there is much ground to cover first to get to the point of understanding on what the epidemiologists paid handsomely to investigate coronaviruses for the decade never bothered to check. No one gets to walk through this in the clear, however, no one. We are all bonded to epochal failures in leadership and institutional cognition. What, for instance, was someone who had worked through COVID's imaginarium early on doing flying a week into March? I too had been infused with a peculiarly American moment, wherein financial desperation meets imperial exceptionalism. I too had to travel for work and nothing was going to happen to me. It's a fetish that working man George Floyd, who two months later died by a cop's knee on a street corner in South Minneapolis I regularly travel, never believed for a moment that he livestreamed himself talking of the daily dangers of being Black. Nor did the series of "mouthy" Black detainees into whom Minneapolis police had paramedics inject ketamine, risking respiratory arrest under the cover of a Hennepin Hospital study. The greatest sources of U.S. wealth are the daily reenactments of the slavery and genocide and environmental decimation on which it was built. From the ritual murders of arrestees to forcing meatpackers back to work during a deadly outbreak, to risk, with COVID attacking our vasculature, blood chokes of their own. As if the country couldn't recognize itself otherwise.

MY DEEPEST GRATITUDE to my co-authors, in alphabetical order: Alex Liebman, David Weisberger, Deborah Wallace, Luca de Crescenzo, Luis Fernando Chaves, Luke Bergmann, Max Ajl, Richard Kock, Rodrick Wallace, Tammi Jonas, and Yaak Pabst. To my colleagues at the Agroecology and Rural Economics Research Corps, Pandemic Research for the People, and Regeneration Midwest— Ann Wolf, Brian Rumsey, Carolyn Betz, Cora Roelofs, Etant Dupain, Graham Christensen, Jasmine Araujo, Jessica Gnad, John Choe,

Graham Christensen, John Gulick, Kaare Melby, Kenichi Okamoto, Kim Williams-Guillen, Laura Paine, Laura Thomas, Meleiza Figueroa, Patrick Kerrigan, Ryan Petteway, Serena Stein, and Tanya Kerssen— thank you for the delightful collaboration.

To my Monthly Review Press compatriots Camila Valle, Erin Clermont, Fred Magdoff, Jamil Jonna, John Bellamy Foster, Martin Paddio, Michael Yates, and Susie Day, thank you for the thousand save-the-days so professionally delivered. To Ben Ehrenreich, Edgar Rivera Colón, Firoze Manji, and Mindy Fullilove, I am humbled by your reports back upon a first reading. The greatest of appreciation to Peter Cury for another excellent cover. And to all my friends online and off, my sincere gratefulness, with grandiose hat tips to Jonathan Latham for our discussions about COVID's origins, Kezia Barker for early editing on the origins of ag disease piece, the Spirit of 1848 listserv for a first-cut critique of our bat cave ideas, and to Anthony Galluzzo and Green Boots for leads to exactly the story hooks I needed.

—JULY 2020

Notes on a Novel Coronavirus

It is not because the Indo-Chinese has discovered a culture of his own that he is in revolt. It is because "quite simply" it was, in more than one way, becoming impossible for him to breathe.

— Frantz Fanon (1952)

A NEW DEADLY CORONAVIRUS, now identified as SARS-CoV-2, related to SARS-1 and MERS and apparently originating in live animal markets in Wuhan, China, is starting to spread worldwide.[6]

Authorities in China have reported 5,974 cases nationwide, 1,000 of them severe.[7] With infections in nearly every province, officials warned SARS-2 appears to be spreading fast out of its epicenter.[8]

The characterization appears supported by initial modeling.[9]

The virus's basic reproduction number, a measure of the number of new cases per infection given no cap on available susceptibles, is clocking in at a healthy 3.11. That means in the face of such momentum, a control campaign must stop up to 75 percent of new infections to reverse the outbreak. The modeling team estimates there are presently over 21,000 cases, reported or not, in Wuhan alone.

Full-genome sequences of the virus meanwhile show few differences between the samples isolated across China.[10] Slower spread for such a fast-evolving RNA virus would be marked by mutations accumulating place-to-place.

The coronavirus is starting to open up theaters overseas. Travelers with SARS-2 have been treated in Australia, France, Hong Kong, Japan, Malaysia, Nepal, Vietnam, Singapore, South Korea, Taiwan, Thailand, and the United States.[11] Local outbreaks are now starting up within six countries.[12]

As the infection is characterized by human-to-human transmission and an apparent two-week incubation period before the sickness hits, the infection will likely continue to spread across the globe. Whether it will be Wuhan everywhere remains an open question.

The virus's final penetrance worldwide will depend on the difference between the rate of infection and the rate of removing infections by recovery or death.[13] If the infection rate far exceeds removal, then the total population infected may approach the whole of humanity. That outcome, however, would likely be marked by large geographic variation brought about by a combination of dead chance and the differences in how countries responded to their outbreak.

Pandemic skeptics aren't so sure of such a scenario. Far fewer patients have been infected and killed by SARS-2 than even the typical seasonal influenza.[14] But the mistake here is in confusing a moment early in an outbreak for a virus's essentialist nature.

Outbreaks are dynamic. Yes, some burn out, including, maybe, SARS-2. It takes the right evolutionary draw and a little luck to beat out chance extirpation. Sometimes enough hosts don't line up to keep transmission going. Other outbreaks explode. Those that make it on the world stage can be game changers, even if they eventually die out. They upend the everyday routines of even a world already in tumult or at war.[15]

The deadliness of any potential pandemic strain is the meat of the matter, of course.

Should the virus prove less infectious or deadly than initially thought, civilization goes on, however many people are killed. The H1N1 (2009) influenza outbreak that worried so many a decade-plus ago proved less virulent than it first seemed. But even that strain penetrated the global population, and quietly killed patients, at magnitudes far beyond these first follow-up dismissals. H1N1 (2009) killed as many as 579,000 people its first year, producing complications in fifteen times more cases than initially projected from lab tests alone.[16]

The danger here is found in humanity's unprecedented connectivity. H1N1 (2009) crossed the Pacific Ocean in nine days, superseding predictions by the most sophisticated models of the global travel network by months. Airline data show a tenfold increase in travel in China just since the SARS-1 epidemic.

Under such widespread percolation, low mortality for a large number of infections can still cause a large number of deaths. If four billion people are infected at a mortality rate of only 2 percent, a death rate less than half that of the 1918 influenza pandemic, eighty million people are killed. And unlike for seasonal influenza, we have neither herd immunity nor a vaccine to slow it down. Even speeded-up development will at best take three months to produce a vaccine for SARS-2, assuming it even works.[17] Scientists successfully produced a vaccine for the H5N2 avian influenza only after the U.S. outbreak ended.[18]

A critical epidemiological parameter will be the relationship between infectivity and when those infected express symptoms. SARS-1 and MERS proved infectious only upon symptoms.[19] If this bears out for SARS-2, we may be in relatively good shape, all things considered. Even without a vaccine or tailored antivirals, we can immediately quarantine the suddenly sick, breaking chains of transmission with nineteenth-century public health.

Sunday, however, China's health minister, Ma Xiaowei, stunned the world announcing that SARS-2 had expressed infectivity before symptoms.[20] It's such a turnabout that infuriated U.S. epidemiologists are demanding access to the data showing the new infectivity. The shock implies researchers stateside expect the virus couldn't possibly be able to evolve outside what they appear to imagine as some public health archetype. If the new infection life history holds true, health authorities are not going to be able to use symptoms to identify newly active cases.

These unknowns—the exact source, infectivity, penetrance, and possible treatments—together explain why epidemiologists and public health officials are worried about SARS-2. Unlike the seasonal influenzas cited by pandemic skeptics, the uncertainty rattles practitioners.

It is the nature of the job to worry, yes. Worry is built into the very probabilities and systemic errors embodied more broadly in the trade. The damage in failing to prepare for an outbreak that proves deadly far exceeds that from the embarrassment of preparing for an outbreak that fails to live up to the hype.[21] But in an era celebrating austerity, few jurisdictions wish to pay for a disaster that is no guarantee, whatever the collateral benefits of precaution or, on the other end of outcomes, the devastating losses associated with a bad gamble.

The choice of how to respond is often entirely out of epidemiologist hands anyway. The national authorities who will make these decisions juggle multiple and often contrary agendas. Stopping even a deadly outbreak isn't always treated as the most important objective.

While authorities stumble about figuring out what to do, the scale of impact can suddenly engage in escape velocity. As SARS-2 itself demonstrated, moving from a single food market to the world stage in a month, the numbers can ramp up so far and fast that an epidemiologist's best effort, their raison d'être, is dealt a lethal blow by facts on the ground.

MY OWN VISCERAL reactions to this disease round have skipped across worry, disappointment, and impatience.

I'm an evolutionary biologist and public health phylogeographer who has worked on various aspects of these new pandemics for twenty-five years, much of my adult life. As I've written elsewhere, with the help of many others, I have tried parlaying a growing understanding of these pathogens, from the genetic sequences of my initial studies up through economic geographies of land use, the political economy of global agriculture, and the epistemology of science.[22]

Clarity can sour a soul. As my social media chimed with queries about SARS-2, my immediate response bordered on pique and exhaustion. What, pray, do you wish me to say? What do you want me to do about this?

In dispensing advice, personal and professional, to legitimately worried friends and colleagues, I made some wrong calls. To one farmer friend's query about traveling abroad, I advised a surgeon's mask, washing hands before all meals, and stop fucking livestock, bro. Darkly ribald humor gets me through stress, but his earnest reply of "Stop fucking livestock?" showed I had missed my mark. Not a good look on my part at all. I apologized. He laughed about it later.

It's an occupational hazard. There is the danger of an existential dread that arises from the political inertia epidemiologists must square off with in preparing the world for a nigh-on irresistible pandemic their constituencies pretend is no bother until it's too late.

If SARS-2 is indeed the Big Bug, and it is not clear yet if that's the case, there is almost nothing to be done at this point. All we can do is batten down the public health hatches and hope the virus kills only a small part of the world's population instead of 90 percent.

Clearly humanity shouldn't start reacting to a pandemic when it's already underway. It's a total dereliction of any notion of forward-thinking theory or practice. And leaders and their learned supporters worldwide identify themselves as Prometheans![23]

As I wrote seven years ago:

I expect it will be a long time before I address an outbreak of human influenza again other than in passing. While an understandable visceral reaction, getting worried at this point in the process is a bit bass-ackwards. The bug, whatever its point of origin, has long left the barn, quite literally.[24]

This century we've already trainspotted novel strains of African swine fever, *Campylobacter*, *Cryptosporidium*, *Cyclospora*, Ebola, *E. coli* O157:H7, foot-and-mouth disease, hepatitis E, *Listeria*, Nipah virus, Q fever, *Salmonella*, *Vibrio*, *Yersinia*, Zika, and a variety of novel influenza A variants, including H1N1 (2009), H1N2v, H3N2v, H5N1, H5N2, H5Nx, H6N1, H7N1, H7N3, H7N7, H7N9, and H9N2.[25]

And near-nothing *real* was done about any of them. Authorities spent a sigh of relief upon each reversal and immediately took the next roll of the epidemiological dice, risking a snake eyes of maximum virulence and transmissibility.

That approach suffers more than a failure of foresight or nerve. However necessary, emergency interventions cleaning up each of these messes can make matters worse.

You see, sources of intervention compete. And, as my colleagues and I argue, emergency criteria are deployed as impositions in Gramscian hegemony to keep us from talking about structural interventions around power and production. Because, don't you know, we're warned, *IT'S AN EMERGENCY RIGHT NOW!*

Atop this game of keep away, the failure to address structural problems can render these very emergency interventions ineffectual. The Allee threshold that prophylaxes and quarantine aim to push pathogen populations below—so that infections may burn out on their own, unable to find new susceptibles—is set by structural causes.

As our team wrote about the Ebola outbreak in West Africa:

> Commoditizing the forest may have lowered the region's ecosystemic threshold to such a point that no emergency intervention can drive the Ebola outbreak low enough to burn out. Novel spillovers suddenly express larger forces of infection. On the other end of the epicurve, a mature outbreak continues to circulate, with the potential to intermittently rebound.
>
> In short, neoliberalism's structural shifts are no mere background on which the emergency of Ebola takes place. The shifts are the emergency as much as the virus itself. . . . Deforestation and intensive agriculture may strip out traditional agroforestry's stochastic friction, which typically keeps the virus from lining up enough transmission.[26]

Despite both an effective vaccine and antivirals, Ebola is presently undergoing its longest recorded outbreak in the Democratic Republic of Congo.[27] What got lost along the way? Where is our biomedical God now? Blaming the Congolese to cover up this failure is an exercise in colonial displacement, washing imperialism's hands of decades of structural adjustment and regime change in the Global North's favor.[28]

Saying there's nothing we can do isn't quite right either, however, even as the complaint about reacting only upon a new disease's attack still stands.

Within any one locale, there is a left program for an outbreak, including organizing neighborhood brigades in mutual aid, demanding any vaccine and antivirals developed be made available at no cost to everyone here and abroad, pirating antivirals and medical supplies, and securing unemployment and health care coverage as the economy tanks during the outbreak.

But that way of thinking and organizing, an integral part of the left's legacy, appears to have left the building for more performative (and discursive) configurations online.[29]

The reactionary bent to disease control left and right has since pivoted me to assisting efforts in anti-capitalist agricultures and conservation. Let's stop the outbreaks we can't handle from emerging in the first place. At this point in my career, with the structural pacing the emergencies, I generally write about infectious diseases in only tangential terms.

STRUCTURAL CAUSES OF DISEASE are themselves a source of debate. For one, questions remain as to SARS-2's origins.

Much initial attention has been placed on a particular wet food market in Wuhan, with an Orientalist preoccupation with strange and unsavory diets, representing both the end of biodiversity the West itself is destroying and a revolting source of dangerous disease:

> The typical market in China has fruits and vegetables, butchered beef, pork and lamb, whole plucked chickens—with heads and beaks attached—and live crabs and fish, spewing water out of churning tanks. Some sell more unusual fare, including live snakes, turtles and cicadas, guinea pigs, bamboo rats, badgers, hedgehogs, otters, palm civets, even wolf cubs.[30]

Said snakes are brandished as both signifier and signified, a literal source of SARS-2 that also harkens to a paradise lost and original sin from a serpent's maw.[31]

There is epidemiological evidence in favor of the hypothesis. Thirty-three of 585 samples at the Wuhan market were found positive for SARS-2, with 31 at the west end of the market where wildlife trading was concentrated.[32]

On the other hand, only 41 percent of these positive samples were found in market streets where the wildlife were housed. A quarter of the original infectees never visited the Wuhan market or appeared directly exposed.[33] The earliest case was identified before the market was hit.[34] Other infected marketers trafficked in hog alone, a livestock species that expresses a common vulnerable molecular receptor, leading one team to hypothesize hog as the putative source for the new coronavirus.

Atop African swine fever, which has killed as many as half of China's hogs this past year, the latter possibility would represent quite the clusterfuck.[35] Such disease convergences are not unheard of, even folding into an intimate reciprocal activation, wherein proteins of each pathogen catalyze each other, facilitating new clinical courses and transmission dynamics for both diseases.[36]

At the same time, Western Sinophobia doesn't absolve Chinese public health.[37] Certainly the anger and disappointment the Chinese public has directed at local and federal authorities for their slow reaction to SARS-2 can't be spun as weaponized xenophobia.[38] And in our wise efforts to keep our foot out of that trap, we may also be missing a critical agroecological symmetry.

Setting aside the culture war, wet markets and "exotic" food are staples in China, as is now industrial production, juxtaposed alongside each other since economic liberalization post-Mao.[39] Indeed, the two food modes may be integrated by way of land use.

Expanding industrial production may push increasingly capitalized wild foods deeper into the last of the primary landscape, dredging out a wider variety of potentially protopandemic pathogens.[40] Periurban loops of growing extent and population density may increase the interface (and spillover) between wild nonhuman populations and newly urbanized rurality.[41]

Worldwide, even the wildest subsistence species are being roped into ag value chains: among them, ostriches, porcupine, crocodiles, fruit bats, and the palm civet, whose partially digested berries now supply the world's most expensive coffee bean.[42] Some wild species are making it onto forks before they are even scientifically identified, including one new short-nosed dogfish found in a Taiwanese market.[43]

All are increasingly treated as food commodities. As nature is stripped place-by-place, species-by-species, what's left over becomes that much more valuable.[44]

Weberian anthropologist Lyle Fearnley pointed out that farmers in China repeatedly manipulate the distinction between wildness and domesticity as an economic signifier, producing new meanings and values attached to their animals, including in response to the very epidemiological alerts issued around their trade.[45] A Marxist might push back that these signifiers emerge out of a context that extends well beyond smallholder control and out onto global circuits of capital.[46]

So while the distinction between factory farms and wet markets isn't unimportant, we may miss their similarities (and dialectical relationships).

The distinctions bleed together by a number of other mechanisms. Many a smallholder worldwide, including in China, is in actuality a contractor, growing out day-old poultry, for instance, for industrial processing.[47] So on a contractor's smallholding along the forest edge, a food animal may catch a pathogen before being shipped back to a processing plant in the outer ring of a major city.

Spreading factory farms, meanwhile, may force increasingly corporatized wild foods companies to trawl deeper into the forest, increasing the likelihood of picking up a new pathogen, while reducing the kind of environmental complexity with which the forest disrupts transmission chains.

Capital weaponizes the resulting disease investigations. Blaming smallholders is now a standard agribusiness crisis management practice, but clearly diseases are a matter of systems of production over time and space and mode, not just specific actors between whom we can juggle blame.[48]

As a class, the coronaviruses appear to straddle these distinctions. While SARS-1 and SARS-2 appeared to have emerged out of wet markets—possible pigs aside—MERS, the other deadly coronavirus, emerged straight out of an industrializing camel sector in the Middle East.[49] It's a path to virulence largely left out of broader scientific discussion about these viruses.

It should change how we think about them. I would recommend we err on the side of viewing disease causality and intervention beyond the biomedical or even ecohealth object and out into the field of ecosocial relationships.[50]

Other ethoses see a different way out. Some researchers recommend we genetically engineer poultry and livestock to be resistant to these diseases.[51] They leave out whether that would still allow these strains to circulate among what would now be asymptomatic food animals before spilling over into decidedly unengineered humans.

Again turning back the clock, a source of my pique, I wrote nine years ago about what efforts at genetically engineering out pathogens miss as matters of first principle:

> Beyond the issue of the affordability of the new frankenchicken, especially for the poorest countries, influenza's success arises in part from its capacity to outwit and outlast such silver bullets. Hypotheses tied to a lucrative model of biology are routinely mistaken for expectations about material reality, expectations are mistaken for projections, and projections for predictions.
>
> One source of vexation is the dimensionality of the problem. There is even among mainstream scholars a dawning realization influenza is more than mere virion or infection; that it respects little of disciplinary boundaries (and business plans) in both their form and content. Pathogens regularly use processes accumulating at one level of biocultural organization to solve problems they face at other levels, including the molecular.[52]

Agribusiness ever turns us toward a techno-utopian future to keep us in a past bounded by capitalist relations. We are spun round and round the very commodity tracks selecting for new diseases in the first place.

THE SECRET THRILL (and open terror) epidemiologists feel during an outbreak is nothing more than defeat disguised as heroism.

Almost the entirety of the profession is presently organized around post hoc duties, much like a stable boy with a shovel following behind the elephants at a circus. Under the neoliberal program, epidemiologists and public health units are funded to clean up the system's mess, while rationalizing even the worst practices that lead to many a deadly pandemic's emergence.[53]

In a commentary on the new coronavirus, one Simon Reid, a professor of communicable disease control at the University of Queensland, instantiates the resulting incoherence.[54] He pings from topic to topic, failing to weave a whole out of his technicist observations. Such folly isn't necessarily a matter of incompetence or malicious intent on Reid's part. It is more a matter of the contradictory obligations of the neoliberal university.

U.S. leftists recently joined swords over the existence of the professional-managerial class or PMC.[55] *Jacobin* social democrats rail at the capitalist PMC they are angling to join in a Sanders administration, while tankies—modern-day Stalinists—claim managers are proletarian too.[56] I'll sidestep the metaphysical debate—how many PMC can dance on an Epipen?—only to observe that whether the PMC theoretically exists in epidemiology, I've met its members in the flesh.[57] They live!

Reid and other institutional epidemiologists are on the hook for cleaning up diseases of neoliberal origins—yes, including out of China—while meting out comforting platitudes that say the system that pays them works. It's a double bind many practitioners choose to live with, nay, prosper from, even should the resulting epidemiologies threaten millions.

Reid here kinda gets the food system and conservation parts of the explanation for SARS-2—and many of its celebrity forerunners in the series of epidemiological reality shows run this century so far. But in introducing this protopandemic, he propositions, to paraphrase, that "This utter horror has a saving grace—hooray!" And, again paraphrasing, it is that "China has been a source of repeated outbreaks, but it, and a WHO now owned by philanthrocapitalism, conducts exemplary biocontrol."[58]

We can reject Sinophobia, offer material support, and still well remember China covered up the SARS-1 outbreak in 2003.[59] Beijing suppressed media and public health reports, allowing that coronavirus to splatter across its own country. Medical authorities one province over from an outbreak didn't know what their patients were suddenly showing up with in the ER. SARS-1 eventually spread across multiple countries as far as Canada and was barely driven to extirpation.

The new century has meanwhile been marked by China's failure or refusal to unpack its near-perfect storm of rice, duck, and industrial poultry and hog production driving multiple novel strains of influenza. It is treated as a price for prosperity.

This is no Chinese exceptionalism, however. The United States and Europe have served as ground zeros for new influenzas as well, recently H5N2 and H5Nx, and their multinationals and neocolonial proxies drove the emergence of Ebola in West Africa and Zika in Brazil.[60] U.S. public health officials covered for agribusiness during the H1N1 (2009) and H5N2 outbreaks.[61]

Perhaps, then, we should refrain from choosing between one of two cycles of capital accumulation: the end of the American cycle or the start of the Chinese one (or, as Reid appears to do, both). At the risk of accusations of third campism, choosing neither is another option.

If we must partake in the Great Game, let's choose an ecosocialism that mends the metabolic rift between ecology and economy, and between the urban and the rural and wilderness, keeping the worst of these pathogens from emerging in the first place.[62] Let's choose international solidarity with everyday people the world over.

Let's realize a creaturely communism far from the Soviet model. Let's braid together a new world system, indigenous liberation, farmer autonomy, strategic rewilding, and place-specific agroecologies that, redefining biosecurity, reintroduce immune firebreaks of widely diverse varieties in livestock, poultry, and crops.

Let's reintroduce natural selection as an ecosystem service and let our livestock and crops reproduce on-site, whereby they can pass along their outbreak-tested immunogenetics to the next generation.[63] Consider the options otherwise.

Maybe I've been unfair to the Reids of the world, who, as a matter of professional obligation, must believe their own contradictions. But, as five hundred years of war and pestilence demonstrate, the sources of capital that many epidemiologists now serve are more than willing to scale mountains made of body bags.

—*MRonline,* JANUARY 29, 2020

"Agribusiness Would Risk Millions of Deaths"

In early March 2020, I was interviewed by Yaak Pabst for Marx21, *the German socialist publication.*[64] *Luca de Crescenzo added the final two questions for an Italian translation a couple weeks later. My answers to these last two, with the U.S. epidemic starting to accelerate, took on a decided shift in tone.*[65]

Yaak Pabst (YP): *How dangerous is the new coronavirus?*
Rob Wallace (RW): It depends on where you are in the timing of your local outbreak of COVID-19: early, peak level, late? How good is your region's public health response? What are your demographics? How old are you? Are you immunologically compromised? What is your underlying health? To ask an undiagnosable possibility, do your immunogenetics, the genetics underlying your immune response, line up with the virus or not?

YP: *So all this fuss about the virus is just scare tactics?*
RW: No, certainly not. At the population level, COVID-19 was clocking in at between 2 and 4 percent case fatality ratio, or CFR, at the start of the outbreak in Wuhan. Outside Wuhan, the CFR appears to

drop off to more like 1 percent and even less, but also appears to spike in spots here and there, including in places in Italy and the United States. Its range doesn't seem much in comparison to, say, SARS-1 at 10 percent, the influenza of 1918 5–20 percent, avian influenza H5N1 60 percent, or Ebola at some points 90 percent. But it certainly exceeds seasonal influenza's 0.1 percent CFR. The danger isn't just a matter of the death rate, however. We have to grapple with what's called penetrance or community attack rate: how much of the global population is penetrated by the outbreak.

YP: *Can you be more specific?*
RW: The global travel network is at record connectivity. With no vaccines or specific antivirals for coronaviruses, nor at this point any herd immunity to the virus, even a strain at only 1 percent mortality can present a considerable danger. With an incubation period of up to two weeks and increasing evidence of some transmission before sickness—before we know people are infected—few places would likely be free of infection. If COVID-19 registers 1 percent fatality in the course of infecting, say, four billion people, that's 40 million dead. A small proportion of a large number can still be a large number.

YP: *These are frightening numbers for an ostensibly less than virulent pathogen.*
RW: Definitely, and we are only at the beginning of the outbreak. It's important to understand that many new infections change over the course of epidemics. Infectivity, virulence, or both may attenuate. On the other hand, other outbreaks ramp up in virulence. The first wave of the influenza pandemic in the spring of 1918 was a relatively mild infection. It was the second and third waves that winter and into 1919 that killed millions.

YP: *But pandemic skeptics argue that far fewer patients have been infected and killed by the coronavirus than by the typical seasonal flu. What do you think about that?*
RW: I would be the first to celebrate if this outbreak proves a dud.

But these efforts to dismiss COVID-19 as a possible danger by citing other deadly diseases, especially influenza, is a rhetorical device to spin concern about the coronavirus as badly placed.

YP: *So the comparison with seasonal flu is limping.*

RW: It makes little sense to compare two pathogens on different parts of their epicurves. Yes, seasonal influenza infects many millions worldwide, killing, by WHO estimates, up to 650,000 people a year. COVID-19, however, is only starting its epidemiological journey. And unlike influenza, we have neither vaccine nor herd immunity to slow infection and protect the most vulnerable populations.

YP: *Even if the comparison is misleading, both diseases belong to viruses, even to a specific group, the RNA viruses. Both can cause disease. Both affect the mouth and throat area and sometimes also the lungs. Both are quite contagious.*

RW: Those are superficial similarities that miss a critical part in comparing two pathogens. We know a lot about influenza's dynamics. We know very little about COVID-19's. They're steeped in unknowns. Indeed, there is much about COVID-19 that is even unknowable until the outbreak plays out fully. At the same time, it is important to understand that it isn't a matter of COVID-19 versus influenza. It's COVID-19 *and* influenza. The emergence of multiple infections capable of going pandemic, attacking populations in combos, should be a front and center worry.

YP: *You have been researching epidemics and their causes for several years. In your book* Big Farms Make Big Flu, *you attempt to draw these connections between industrial farming practices, organic farming, and viral epidemiology. What are your insights?*

RW: The real danger of each new outbreak is the failure, or better put, the expedient refusal to grasp that each new COVID-19 is no isolated incident. The increased occurrence of novel viruses is closely linked to food production and the profitability of multinational corporations. Anyone who aims to understand why viruses are becoming more dangerous must investigate the industrial model of agriculture and, more specifically, livestock production. At present, few governments, and few scientists, are prepared to do so. Quite the contrary.

When the new outbreaks spring up, governments, the media, and even most of the medical establishment are so focused on each separate emergency that they dismiss the structural causes that are driving multiple marginalized pathogens into sudden global celebrity, one after the other.

YP: *Who is to blame?*
RW: I said industrial agriculture, but there's a larger scope to it. Capital is spearheading land grabs into the last of primary forest and smallholder farmland worldwide. These investments drive the deforestation and development leading to disease emergence. The functional diversity and complexity these huge tracts of land represent are being streamlined in such a way that previously boxed-in pathogens are spilling over into local livestock and human communities. In short, capital centers, places such as London, New York, and Hong Kong, should be considered our primary disease hotspots.

YP: *For which diseases is this the case?*
RW: There are no capital-free pathogens at this point. Even the most remote are affected, if distally. Ebola, Zika, the coronaviruses, yellow fever again, a variety of avian influenzas, and African swine fever in hogs are among the many pathogens making their way out of the most remote hinterlands into periurban loops, regional capitals, and ultimately onto the global travel network. From fruit bats in the Congo to killing Miami sunbathers in a few weeks' time.

YP: *What is the role of multinational companies in this process?*
RW: Planet Earth is largely Planet Farm at this point, in both biomass and land used. Agribusiness is aiming to corner the food market. The near-entirety of the neoliberal project is organized around supporting efforts by companies based in the more advanced industrialized countries to steal the land and resources of weaker countries. As a

result, many of those new pathogens previously held in check by long-evolved forest ecologies are being sprung free, threatening the whole world.

YP: *What effects do the production methods of agribusinesses have on this?*
RW: The capital-led agriculture that replaces more natural ecologies offers the exact means by which pathogens can evolve the most virulent and infectious phenotypes. You couldn't design a better system to breed deadly diseases.

YP: *How so?*
RW: Growing genetic monocultures of domestic animals removes whatever immune firebreaks may be available to slow down transmission. Larger population sizes and densities facilitate greater rates of transmission. Such crowded conditions depress immune response. High throughput, a part of any industrial production, provides a continually renewed supply of susceptibles, the fuel for the evolution of virulence. In other words, agribusiness is so focused on profits that selecting for a virus that might kill a billion people is treated as a worthy risk.

YP: *What!?*
RW: These companies can just externalize the costs of their epidemiologically dangerous operations on everyone else. From the animals themselves to consumers, farmworkers, local environments, and governments across jurisdictions. The damages are so extensive that if we were to return those costs to company balance sheets, agribusiness as we know it would be ended forever. No company could support the costs of the damage it imposes.

YP: *In many media, it is claimed that the starting point of the coronavirus was an "exotic food market" in Wuhan. Is this description true?*
RW: Yes and no. There are spatial clues in favor of the notion. Contact tracing linked infections back to the Huanan Wholesale Sea Food

Market in Wuhan, where wild animals were sold. Environmental sampling does appear to pinpoint the west end of the market where wild animals were held.

But how far back and how widely should we investigate? When exactly did the emergency really begin? The focus on the market misses the origins of wild agriculture out in the hinterlands and its increasing capitalization. Globally, and in China, wild food is becoming more formalized as an economic sector. But its relationship with industrial agriculture extends beyond merely sharing the same moneybags. As industrial production—hog, poultry, and the like—expands into primary forest, it places pressure on wild food operators to dredge further into the forest for source populations, increasing the interface with, and spillover of, new pathogens, including COVID-19.

YP: *COVID-19 is not the first virus to develop in China that the government tried to cover up.*
RW: Yes, but this is no Chinese exceptionalism. The United States and Europe have served as ground zeros for new influenzas as well, recently H5N2 and H5Nx, and their multinationals and neocolonial proxies drove the emergence of Ebola in West Africa and Zika in Brazil. U.S. public health officials covered for agribusiness during the H1N1 (2009) and H5N2 outbreaks.

YP: *The World Health Organization (WHO) has now declared a "health emergency of international concern." Is this step correct?*
RW: Yes. The danger of such a pathogen is that health authorities do not have a handle on the statistical risk distribution. We have no idea how the pathogen may respond. We went from an outbreak in a market to infections splattered across the world in a matter of weeks. The pathogen could just burn out. That would be great. But we don't know. Better preparation would better the odds of undercutting the pathogen's escape velocity.

The WHO's declaration is also part of what I call pandemic theater. International organizations have died in the face of inaction. The League of Nations comes to mind. The UN group of organizations is always worried about its relevance, power, and funding. But such actionism can also converge on the actual preparation and prevention the world needs to disrupt COVID-19's chains of transmission.

YP: *The neoliberal restructuring of the health care system has worsened both the research and the general care of patients, for example, in hospitals. What difference could a better funded health care system make to fight the virus?*

RW: There's the terrible but telling story of the Miami medical device company employee who upon returning from China with flu-like symptoms did the righteous thing by his family and community and demanded that a local hospital test him for COVID-19. He worried that his minimal Obamacare option wouldn't cover the tests. He was right. He was suddenly on the hook for US$3,270. An American demand might be that an emergency order be passed that stipulates that during a pandemic outbreak, all outstanding medical bills related to testing for infection and for treatment following a positive test would be paid for by the federal government. We want to encourage people to seek help, after all, rather than hide away—and infect others—because they can't afford treatment. The obvious solution is a national health service—fully staffed and equipped to handle such community-wide emergencies—so that such a ridiculous problem as discouraging community cooperation would never arise.

YP: *As soon as the virus is discovered in one country, governments everywhere react with authoritarian and punitive measures, such as a compulsory quarantine of entire areas of land and cities. Are such drastic measures justified?*

RW: Using an outbreak to beta-test the latest in autocratic control post-outbreak is disaster capitalism gone off the rails. In terms of public health, I would err on the side of trust and compassion, which are important epidemiological variables. Without either, jurisdictions lose their populations' support. A sense of solidarity and common respect is a critical part of eliciting the cooperation we need to survive such threats together. Self-quarantines with the proper support—check-ins by trained neighborhood brigades, food supply trucks going door-to-door, work release and unemployment insurance—can elicit that kind of cooperation, that we are all in this together.

YP: *As you may know, in Germany with the AfD we have a de facto Nazi party with 94 seats in parliament. The hard Nazi Right and other groups in association with AfD politicians use the Corona crisis for their agitation. They spread (false) reports about the virus and demand more authoritarian measures from the government: Restrict flights and entry stops for migrants, border closures, and forced quarantine . . .*

RW: Travel bans and border closures are demands with which the radical right wants to racialize what are now global diseases. This is, of course, nonsense. At this point, given the virus is already on its way to spreading everywhere, the sensible thing to do is to work on developing the kind of public health resilience in which it doesn't matter who shows up with an infection, we have the means to treat and cure them. Of course, stop stealing people's land abroad and driving the exoduses in the first place, and we can keep the pathogens from emerging in the first place.

YP: *What would be sustainable changes?*

RW: In order to reduce the emergence of new virus outbreaks, food production has to change radically. Farmer autonomy and a strong public sector can curb environmental ratchets and runaway infections. Introduce varieties of stock and crops—and strategic rewilding—at both the farm and regional levels. Permit food animals to reproduce on-site to pass on tested immunities. Connect just production with just circulation. Subsidize price supports and consumer purchasing programs supporting agroecological production. Defend these experiments from both the compulsions that neoliberal economics impose upon individuals and communities alike and the threat of capital-led State repression.

YP: *What should socialists call for in the face of the increasing dynamics of disease outbreaks?*

RW: Agribusiness as a mode of social reproduction must be ended for good, if only as a matter of public health. Highly capitalized production of food depends on practices that endanger the entirety of humanity, in this case helping unleash a new deadly pandemic. We should demand that food systems be socialized in such a way that pathogens this dangerous are kept from emerging in the first place. This will require reintegrating food production into the needs of rural communities first. That will require agroecological practices that protect the environment and farmers as they grow our food. Big picture, we must heal the metabolic rifts separating our ecologies from our economies. In short, we have a planet to win.

Luca de Crescenzo (LdC): *I would like you to add a comment about the recent proposal of the UK authorities not to take drastic measures to contain the virus and to bet on the development of the herd immunity instead. You wrote: "This is a failure that pretends to be a solution." Can you explain that?*

RW: The Tories are asserting joining the United States in effectively denying health care is the best active cure. The government is looking at parlaying its late response into letting COVID-19 work through the population to produce the herd immunity it says will protect the most vulnerable.

This is the utter opposite of "do no harm," as the doctor's oath goes. This is, let's do maximum damage.

Herd immunity is treated in epidemiological circles as at best a dirty collateral benefit of an outbreak. Enough people carry antibodies from the last outbreak to keep the susceptible population low enough that no new infection could support itself, protecting even those who haven't been previously exposed. It's often no more than a passing effect, however, if the pathogen in question evolves out from underneath the population blanket.

We do better in inducing such immunity by campaigns in vaccination. Typically, such an effect requires a wide majority of people vaccinated to work. Which, outside market failures in producing vaccines, is routinely no problem as *nearly no one dies from them*.

Given the trail of dead in a pandemic, no public health system would actively seek out such a post hoc epiphenomenon as an instrumental objective. No government charged with protecting a population's very lives would allow such a pathogen to run unimpeded, whatever hand-waving is made about "delaying" spread, as if a government already a step behind in responding can exercise such magical control. A campaign of active neglect would kill hundreds of thousands of the very vulnerable that the Tories claim they wish to protect.

But destroying the village to save it is the core premise of a state of the most virulent class character. It's the sign of an exhausted empire that, unable to follow China and other countries in putting up a fight, pretends, as I wrote, that its failures are exactly the solution.

LdC: *In Italy, despite the quarantine and apart from the few who are working from home, a lot of workers still go to work everyday. Many shops are closed but most of the factories are open, even those which don't produce necessary goods. Recently, the trade unions and the federation of the Italian employers have reached an agreement about safe and secure measures at the workplace, which gives the companies only "recommendations" about distance, cleanness, use of masks, without much specification. There are strong reasons to believe they will not be respected. What's your take on that? Is workers' strength an epidemiological variable?*

RW: Working people are treated as cannon fodder. Not only on the battlefield, but back home. Here you have a virus ripping through the Italian population at a rate that exceeds that of the pace it went through China, and capital is pretending it is business—their business—as usual. Negotiating a détente that permits this work to continue without biolab-level precautions is destructive to both workers' standing—you're signaling you'll eat any bowl of shit they serve up—and to the very health of the nation.

If not for your unions' very legitimacy, then for your very lives, and those of your most vulnerable coworkers and community members— shut those factories down! Italy's spike in cases is so dizzying that self-quarantine and negotiated working conditions won't be enough to quash the outbreak. COVID-19 is too infectious and under a medical gridlock too deadly for half-measures. Italy is being invaded by a virus that is kicking the country's ass, with streetfighting door-to-door and home-to-home.

What I'm getting at is that Italy needs to snap the fuck out of it already!

Yes, workers routinely hold up the sky during dark and dangerous days, including during a deadly outbreak. But if the work isn't a matter of the day-to-day operations required during communal quarantine, shut it down. As in countries around the world, the government must then be held responsible for covering the salaries of the workers who have walked off the job in service of the nation's public health.

It's not my call, and my own country is totally botching its response to the pandemic, but should capital resist such efforts to protect the lives of millions, then working Italians, as working people elsewhere, should consider tapping into their proud history of labor militancy and find a means by which to wrestle operative command from the greedy and incompetent. If factories producing non-essential goods are still running, that means management and the moneybags behind them don't give a fuck about you. Even now the chief financial officer upstairs is proving himself more than happy to fold dead workers into the costs of production if he can get away with it.

It wouldn't be the first time the people of the region pushed back during an outbreak. Historian Sheldon Watts noted one unexpected reversal in early disaster capitalism:

> In their rush to save themselves [from plague] by flight, Florentine magistrates worried that the common people left behind would seize control of the city; the fear was perhaps justified. In the summer of 1378 when factional disputes temporarily immobilized the Florentine elite, rebellious woolworkers won control of the government and remained in power for several months.

Several months today might save many thousands of lives. With many countries ten days out from finding themselves in Italy's predicament, working Italians can offer an example for the rest of the world that everyday people's lives matter more than somebody else's profit.

—MARX21, MARCH 11, 2020
UNEVEN EARTH, MARCH 16, 2020

COVID-19 and Circuits of Capital

SANDERSON: Because the city polluted its own water supply, Lower Manhattan needed to find another water source, which led Aaron Burr to form the Manhattan Company. The company charter included a provision that allowed Burr to use most of the assets for something besides water. So he formed a bank, which today is JPMorgan Chase.

KIMMELMAN: Which was Burr's real ambition. He, I think, argued for the water company after the city suffered an outbreak of yellow fever. Then the company built a system so poor it provoked a series of cholera outbreaks.
—Michael Kimmelman and Eric Sanderson (2020)

I co-authored the following piece with human geographer Alex Liebman, disease ecologist Luis Fernando Chaves, and mathematical epidemiologist Rodrick Wallace for Monthly Review, *with perspicacious comment from evolutionary biologist Kenichi Okamoto.*

Calculation

COVID-19, THE ILLNESS caused by coronavirus SARS-CoV-2, the second severe acute respiratory syndrome virus since 2002, is now officially a pandemic. As of late March, whole cities are sheltered in place and, one by one, hospitals are lighting up in medical gridlock brought about by surges in patients.

China, its initial outbreak in contraction, presently breathes easier.[66] South Korea and Singapore as well. Europe, especially Italy and Spain, but increasingly other countries, already bends under the weight of deaths still early in the outbreak. Latin America and Africa are only now beginning to accumulate cases, some countries preparing better than others. In the United States, a bellwether if only as the richest country in the history of the world, the near future looks bleak. The outbreak is not slated to peak stateside until May, and already health care workers and hospital visitors are fistfighting over access to the dwindling supply of personal protection equipment.[67] Nurses, to whom the Centers for Disease Control and Prevention (CDC) has appallingly recommended using bandanas and scarves as masks, have already declared that "the system is doomed."[68]

The U.S. administration meanwhile continues to outbid individual states for basic medical equipment that it refused to purchase for them in the first place. It has also announced a border crackdown as a public health intervention, while the virus rages on ill-addressed inside the country.[69]

An epidemiology team at Imperial College projected that the best campaign in *mitigation*—flattening the plotted curve of accumulating cases by quarantining detected cases and socially distancing the elderly—would still leave the United States with 1.1 million dead and a case burden eight times the country's total critical care beds.[70] Disease *suppression*, looking to end the outbreak, would take public health further into a China-style patient and family member quarantine and community-wide distancing, including closing down institutions. That would bring the United States down to a projected range of around 200,000 deaths.

The Imperial College group estimates a successful campaign in suppression would have to be pursued for at least eighteen months, carrying an overhead in economic contraction and decay in community services. The team proposed balancing the demands of disease control and economy by toggling in and out of community quarantine, as triggered by a set level of critical care beds filled.

Other modelers have pushed back. A group led by Nassim Taleb of *Black Swan* fame declares the Imperial College model fails to include contact tracing and door-to-door monitoring.[71] Their counterpoint misses that the outbreak has broken past many governments' willingness to engage that kind of *cordon sanitaire*. It will not be until the outbreak begins its decline that many countries will view such measures, hopefully with a functional and accurate test, as appropriate. As one droll Tweeter put it: "Coronavirus is too radical. America needs a more moderate virus that we can respond to incrementally."[72]

The Taleb group notes the Imperial team's refusal to investigate under what conditions the virus can be driven to extinction. Such extirpation does not mean zero cases, but enough isolation that single cases are unlikely to produce new chains of infection. Only 5 percent of susceptibles in contact with a case in China were subsequently infected. In effect, the Taleb team favors China's suppression program, going all out, fast enough to drive the outbreak to extinction without getting into a marathon dance toggling between disease control and ensuring the economy no labor shortage. In other words, China's strict (and resource-intensive) approach frees its population from the months-long—or even years-long—sequestration in which the Imperial team recommends other countries partake.

Mathematical epidemiologist Rodrick Wallace, one of us, overturns the modeling table entirely. The modeling of emergencies, however necessary, misses when and where to begin. Structural causes are as much part of the emergency. Including them helps us figure out how best to respond moving forward, beyond just restarting the economy that produced the damage. "If firefighters are given sufficient resources," writes Wallace,

> under normal conditions, most fires, most often, can be contained with minimal casualties and property destruction. However, that containment is critically dependent on a far less romantic, but no less heroic enterprise, the persistent, ongoing,

regulatory efforts that limit building hazard through code development, and enforcement, and that also ensure firefighting, sanitation, and building preservation resources are supplied to all at needed levels.

Context counts for pandemic infection, and current political structures that allow multinational agricultural enterprises to privatize profits while externalizing and socializing costs must become subject to "code enforcement" that reinternalizes those costs if truly mass-fatal pandemic disease is to be avoided in the near future.[73]

The failure to prepare for and react to the outbreak did not just start in December when countries around the world failed to respond once COVID-19 spilled out of Wuhan. In the United States, for instance, it did not start when Donald Trump dismantled his national security team's pandemic preparation team or left 700 CDC positions unfilled.[74] Nor did it start when feds failed to act on the results of a 2017 pandemic simulation showing the country was unprepared.[75] Nor when, as stated in a Reuters headline, the United States "axed CDC expert job in China months before virus outbreak," although missing the early direct contact from a U.S. expert on the ground in China certainly weakened the U.S. response. Nor did it start with the unfortunate decision not to use the already available test kits provided by the World Health Organization. Together, the delays in early information and the total miss in testing will undoubtedly be responsible for many, probably thousands, of lost lives.[76]

The failures were actually programmed decades ago as the shared commons of public health were simultaneously neglected and monetized.[77] A country captured by a regimen of individualized, just-in-time epidemiology—an utter contradiction—with barely enough hospital beds and equipment for normal operations, is by definition unable to marshal the resources necessary to pursue a China brand of suppression.

Following up the Taleb team's point about modeling strategies in more explicitly political terms, disease ecologist Luis Fernando Chaves, another co-author of this article, references dialectical biologists Richard Levins and Richard Lewontin to concur that "letting the numbers speak" only masks all the assumptions folded in beforehand.[78] Models such as the Imperial study explicitly limit the scope of analysis to narrowly tailored questions framed within the dominant social order. By design, they fail to capture the broader market forces driving outbreaks and the political decisions underlying interventions.

Consciously or not, the resulting projections set securing health for all in second place, including the many thousands of the most vulnerable who would be killed should a country toggle between disease control and the economy. The Foucauldian vision of a state acting on a population in its own interests only represents an update, albeit a more benign one, of the Malthusian push for herd immunity that Britain's Tory government and now the Netherlands proposed—letting the virus burn through the population unimpeded.[79] There is little evidence beyond an ideological hope that herd immunity would guarantee stopping the outbreak. The virus may readily evolve out from underneath the population's immune blanket.

Intervention

What should be done instead? First, we need to grasp that, in responding to the emergency the right way, we will still be engaging in both necessity and danger.

We need to nationalize hospitals as Spain did in response to the outbreak.[80] We need to supercharge testing in volume and turnaround as Senegal has.[81] We need to socialize pharmaceuticals.[82] We need to enforce maximum protections for medical staff to slow staff decay. We must secure the right to repair for ventilators and other medical machinery.[83] We need to start mass-producing cocktails of antivirals such as remdesivir and old-school antimalarial chloroquine (and any other drugs that appear promising) while we conduct clinical trials testing whether they work beyond the laboratory.[84] A planning system should be implemented to (1) force companies to produce the needed ventilators and personal protection equipment required by health care workers and (2) prioritize allocation to places with the greatest needs.

We must establish a massive pandemic corps to provide the workforce—from research to care—that approaches the order of demand the virus (and any other pathogen to come) is placing on us. Match the caseload with the number of critical care beds, staffing, and equipment necessary so that suppression can bridge the present numbers gap. In other words, we cannot accept the idea of merely surviving COVID-19's ongoing air attack only to return later to contact tracing and case isolation to drive the outbreak below its threshold. We must hire enough people to identify COVID-19 home-by-home right now and equip them with the needed protective gear, such as adequate masks. Along the way, we need to suspend a society organized around expropriation, from landlords up through sanctions on other countries, so that people can survive both the disease and its cure.

Until such a program can be implemented, however, the greater populace is left largely abandoned. Even as continued pressure must be brought to bear on recalcitrant governments, in the spirit of a largely lost tradition in proletarian organizing going back 150 years, everyday people who are able should join emerging mutual aid groups and neighborhood brigades.[85] Professional public health staff that unions can spare should train these groups to keep acts of kindness from spreading the virus.

The insistence that we fold the virus's structural origins into emergency planning offers us a key to parlaying every step forward into protecting people before profits.

One of many perils lies in normalizing the "batshit crazy" presently underway, a serendipitous characterization given the syndrome that patients suffer—proverbial "bat shit" in the lungs. We need to retain the shock we received when we learned another SARS virus emerged out of its wildlife refugia and in a matter of eight weeks splattered itself across humanity.[86] The virus emerged at one terminus of a regional supply line in "exotic foods," successfully setting off a human-to-human chain of infections at the other end in Wuhan, China.[87] From there, the outbreak both diffused locally and hopped onto planes and trains, spreading out across the globe through a web structured by travel connections and down a hierarchy from larger to smaller cities.[88]

Other than describing the wild food market in the typical Orientalism, little effort has been expended on the most obvious of questions. How did the "exotic food" sector arrive at a standing where it could sell its wares alongside more traditional livestock in the largest market in Wuhan? The animals were not being sold off the back of a truck or in an alleyway. Think of the permits and payments (and deregulation thereof) involved.[89] Well beyond fisheries, worldwide wild food is an increasingly formalized sector, evermore capitalized by the same sources backing industrial production.[90] Although nowhere near similar in the magnitude of output, the distinction is now more opaque.

The overlapping economic geography extends back from the Wuhan market to the hinterlands where exotic and traditional foods are raised by operations bordering the edge of a contracting wilderness.[91] As industrial production encroaches on the last of the forest, wild food operations must cut further in to raise their delicacies or raid the last stands. As a result, the most exotic of pathogens, in this case bat-hosted SARS-2, find their way onto a truck, whether in food animals or the labor tending them, shotgun from one end of a lengthening periurban circuit to the other before hitting the world stage.[92]

Infiltration

The connection bears elaboration, as much in helping us plan forward during this outbreak as in understanding how humanity maneuvered itself into such a trap.

Some pathogens emerge right out of centers of production. Foodborne bacteria such as *Salmonella* and *Campylobacter* come to mind. But many like COVID-19 originate on the frontiers of capitalist production. Indeed, a majority of the novel human pathogens that emerge from animals spill over from wildlife to local human communities (before the more successful ones spread to the rest of the world).[93]

A number of luminaries in the field of ecohealth, some funded in part by Colgate-Palmolive and Johnson & Johnson, companies driving the bleeding edge of agribusiness-led deforestation, produced a global map based on previous outbreaks back to 1940 intimating where new pathogens are likely to emerge moving forward.[94] The warmer the color on the map, the more likely a new pathogen should emerge there. But in confusing such *absolute geographies*, the team's map—red hot in China, India, Indonesia, and parts of Latin America and Africa—missed a critical point. Focusing on outbreak zones ignores the relations shared by global economic actors that shape epidemiologies.[95] The capitalist interests backing development- and production-induced changes in land use and disease emergence in underdeveloped parts of the globe reward efforts that pin responsibility for outbreaks on indigenous populations and their so-deemed "dirty" cultural practices.[96] Prepping bushmeat and home burials are two practices blamed for the emergence of new pathogens. Plotting *relational geographies*, in contrast, suddenly turns New York, London, and Hong Kong, key sources of global capital, into three of the world's worst disease hotspots instead.

Meanwhile, outbreak zones are no longer organized under traditional polities. Unequal ecological exchange—redirecting the worst damage from industrial agriculture to the Global South—has moved from stripping localities of resources by state-led imperialism and into new complexes across scale and commodity.[97] Agribusiness is reconfiguring its extractivist operations into spatially discontinuous networks across territories of differing scales.[98] A series of multinational-based "Soybean Republics," for instance, now ranges across Bolivia, Paraguay, Argentina, and Brazil. The new geography is embodied by changes in company management structure, capitalization, subcontracting, supply chain substitutions, leasing, and transnational land pooling.[99] In straddling national borders, these "commodity countries," flexibly embedded across ecologies and political borders, are producing new epidemiologies along the way.[100]

For instance, despite a general shift in population from commoditized rural areas to urban slums that continues today across the globe, the rural-urban divide driving much of the discussion around disease emergence misses rural-destined labor and the rapid growth of rural towns into periurban *desakotas* (city villages) or *Zwischenstadt* (in-between cities). Mike Davis and others have identified how these newly urbanizing landscapes act as both local markets and regional hubs for global agricultural commodities passing through.[101] Some such regions have even gone "post-agricultural."[102] As a result, forest disease dynamics, the pathogens' primeval sources, are no longer constrained to the hinterlands alone. Their associated epidemiologies have themselves turned relational, felt across time and space. A SARS can suddenly find itself spilling over into humans in the big city a short time out of its bat cave.

Ecosystems in which such "wild" viruses were in part controlled by the complexities of the tropical forest are being drastically streamlined by capital-led deforestation and, at the other end of periurban development, by deficits in public health and environmental sanitation.[103] While many sylvatic pathogens are dying off with their host species as a result, a subset of infections that once burned out relatively quickly in the forest, if only by an irregular rate of encountering their typical host species, are now propagating across susceptible human populations whose vulnerability to infection is in cities often exacerbated by austerity programs and corrupted regulation. Even in the face of efficacious vaccines, the resulting outbreaks are characterized by greater extent, duration, and momentum.[104] What were once local spillovers are now epidemics trawling their way through global webs of travel and trade.

By this parallax effect—by a change in the environmental background alone—old standards such as Ebola, Zika, malaria, and yellow fever, evolving comparatively little, have all made sharp turns into regional threats.[105] They have suddenly moved from spilling over into remote villagers now and again to infecting thousands in capital cities. In something of the other ecological direction, even wild animals, routinely longtime disease reservoirs, are suffering blowback. Their populations fragmented by deforestation, native New World monkeys susceptible to wild-type yellow fever, to which they had been exposed for at least a hundred years, are losing their herd immunity and dying in the hundreds of thousands.[106]

Expansion

If by its global expansion alone, commodity agriculture serves as both propulsion for and nexus through which pathogens of diverse origins migrate from the most remote reservoirs to the most international of population centers.[107] It is here, and along the way, where novel pathogens infiltrate agriculture's gated communities. The lengthier the associated supply chains and the greater the extent of adjunct deforestation, the more diverse (and exotic) the zoonotic pathogens that enter the food chain. Among recent emergent and reemergent farm and foodborne pathogens, originating from across the anthropogenic domain, are African swine fever, *Campylobacter*, *Cryptosporidium*, *Cyclospora*, Ebola Reston, *E. coli* O157:H7, foot-and-mouth disease, hepatitis E, *Listeria*, Nipah virus, Q fever, *Salmonella*, *Vibrio*, *Yersinia*, and a variety of novel influenza variants, including H1N1 (2009), H1N2v, H3N2v, H5N1, H5N2, H5Nx, H6N1, H7N1, H7N3, H7N7, H7N9, and H9N2.[108]

However unintended, the entirety of the production line is organized around practices that accelerate the evolution of pathogen virulence and subsequent transmission.[109] Growing genetic monocultures—food animals and plants with nearly identical genomes—removes immune firebreaks that in more diverse populations slow down transmission.[110] Pathogens now can just quickly evolve around the commonplace host-immune genotypes. Meanwhile, crowded conditions depress immune response.[111] Larger farm animal population sizes and factory farm densities facilitate greater transmission and recurrent infection.[112] High throughput, a part of any industrial production, provides a continually renewed supply of susceptibles at barn, farm, and regional levels, removing the cap on the evolution of pathogen deadliness.[113] Housing a lot of animals together rewards those strains that can burn through them best. Decreasing the age of slaughter—to six weeks in chickens—is likely to select for pathogens able to survive more robust immune systems.[114] Lengthening the geographic extent of live animal trade and export has increased the diversity of genomic segments that their associated pathogens exchange, increasing the rate at which disease agents explore their evolutionary possibilities.[115]

While pathogen evolution rockets forward in all these ways, there is little to no intervention, even at the industry's own demand, save what is required to rescue any single quarter's fiscal margins from the sudden emergency of an outbreak.[116] The trend tends toward fewer government inspections of farms and processing plants, legislation *against* government surveillance and activist exposé, and legislation against even reporting on the specifics of deadly outbreaks in media outlets. Despite recent court victories against pesticide and hog pollution, the private command of production remains entirely focused on profit. The damages caused by the outbreaks that result are externalized to livestock, crops, wildlife, workers, local and national governments, public health systems, and alternate agrosystems abroad as a matter of national priority. In the United States, the CDC reports foodborne outbreaks are expanding in the numbers of states impacted and people infected.[117]

That is, capital's alienation is parsing out in pathogens' favor. While the public interest is filtered out at the farm and food factory gate, pathogens bleed past the biosecurity that industry is willing to pay for and back out to the public. Everyday production represents a lucrative moral hazard eating through our shared health commons.

Liberation

There is a telling irony in New York, one of the largest cities in the world, sheltering in place against COVID-19, a hemisphere away from the virus's origins. Millions of New Yorkers are hiding out in housing stock overseen until recently by one Alicia Glen, until 2018 the city's deputy mayor for housing and economic development.[118] Glen is a former Goldman Sachs executive who oversaw the investment company's Urban Investment Group, which finances projects in the kinds of communities the firm's other units help redline.[119]

Glen, of course, is not in any way personally to blame for the out-break, but she is a symbol of a connection that hits closer to home. Three years before the city hired her, upon a housing crisis and Great Recession in part of its own making, her former employer, along with JPMorgan, Bank of America, Citigroup, Wells Fargo, and Morgan Stanley, took 63 percent of the resulting federal emergency loan financing.[120] Goldman Sachs, cleared of overhead, moved to diversifying its holdings out of the crisis. Goldman Sachs took 60 percent stock in Shuanghui Investment and Development, part of the giant Chinese agribusiness that bought U.S.-based Smithfield Foods, the largest hog producer in the world.[121] For $300 million, it also scored out-and-out ownership of ten poultry farms in Fujian and Hunan, one province over from Wuhan and well within the city's wild foods catchment.[122] It invested up to another $300 million alongside Deutsche Bank in hog raising in the same provinces.[123]

The relational geographies of food production explored above have circled all the way back. There is the pandemic presently sickening Glen's constituencies, apartment-to-apartment across New York, presently the largest U.S. COVID-19 epicenter. But we need also to acknowledge that the loop of causes of the outbreak in part extended out from New York to begin with, however minor in this instance Goldman Sachs's investment may be for a system the size of China's agriculture.

Nationalistic finger pointing, from Trump's racist "China virus" and across the liberal continuum, obscures the interlocking global directorates of state and capital.[124] "Enemy brothers," Karl Marx described them.[125] The death and damage borne by working people on the battlefield, in the economy, and now on their couches fighting to catch their breath manifest both the competition among elites maneuvering for dwindling natural resources and the means shared in dividing and conquering the mass of humanity caught in the gears of these machinations.

Indeed, a pandemic that arises out of the capitalist mode of production and that the state is expected to manage on one end can offer an opportunity from which the system's managers and beneficiaries can prosper on the other. In mid-February, five U.S. senators and twenty House members dumped millions of dollars in personally held stock in industries likely to be damaged in the oncoming pandemic.[126] The politicos based their insider trading on nonpublic intelligence, even as some of the representatives continued to publicly repeat regime missives that the pandemic served no such threat.

Beyond such crass smash-and-grabs, the corruption stateside is systemic, a marker of the end of the U.S. cycle of accumulation when capital cashes out.

There is something comparatively anachronistic in efforts to keep the spout on even if organized around reifying finance over the reality of the primary ecologies (and related epidemiologies) on which it is based. For Goldman Sachs itself, the pandemic, as crises before, offers "room to grow":

> We share the optimism of the various vaccine experts and researchers at biotech companies based on the good progress that has been made on various therapies and vaccines so far. We believe that fear will abate at the first significant evidence of such progress. . . .
>
> Trying to trade to a possible downside target when the year-end target is substantially higher is appropriate for day traders, momentum followers, and some hedge fund managers, but not for long-term investors. Of equal importance, there is no guarantee that the market reaches the lower levels that may be used as justification for selling today. On the other hand, we are more confident that the market will eventually reach the higher target given the resiliency and preeminence of the US economy.
>
> And finally, we actually think that current levels provide an opportunity to slowly add to the risk levels of a portfolio. For those who may be sitting on excess cash and have staying power with the right strategic asset allocation, this is the time to start incrementally adding to S&P equities.[127]

Appalled by the ongoing carnage, people the world over draw

different conclusions.[128] The circuits of capital and production that pathogens mark like radioactive tags one after the other are thought unconscionable.

How to characterize such systems beyond, as we did above, the episodic and circumstantial? Our group is in the midst of deriving a model that outstrips efforts by the modern colonial medicine found in ecohealth and One Health that continues to blame the indigenous and local smallholders for the deforestation that leads to the emergence of deadly diseases.[129]

Our general theory of neoliberal disease emergence, including, yes, in China, combines:

- global circuits of capital;
- deployment of said capital destroying regional environmental complexity that keeps virulent pathogen population growth in check;
- the resulting increases in the rates and taxonomic breadth of spillover events;
- the expanding periurban commodity circuits shipping these newly spilled-over pathogens in livestock and labor from the deepest hinterland to regional cities;
- the growing global travel (and livestock trade) networks that deliver the pathogens from said cities to the rest of the world in record time;
- the ways these networks lower transmission friction, selecting for the evolution of greater pathogen deadliness in both livestock and people;
- and, among other impositions, the dearth of reproduction on-site in industrial livestock, removing natural selection as an ecosystem service that provides real-time (and nearly free) disease protection.

The underlying operative premise is that the cause of COVID-19 and other such pathogens is not found only in the object of any one infectious agent or its clinical course, but also in the field of ecosystemic relations that capital and other structural causes have pinned back to their own advantage.[130] The wide variety of pathogens, representing different taxa, source hosts, modes of transmission, clinical courses, and epidemiological outcomes, have all the earmarks that send us running wild-eyed to our search engines upon each outbreak, and mark different parts and pathways along the same kinds of circuits of land use and value accumulation.

A general program of intervention runs in parallel far beyond a particular virus.

To avoid the worst outcomes here on out, *disalienation* offers the next great human transition: abandoning settler ideologies, reintroducing humanity back into Earth's cycles of regeneration, and rediscovering our sense of individuation in multitudes beyond capital and the state.[131] However, economism, the belief that all causes are economic alone, will not be liberation enough. Global capitalism is a many-headed hydra, appropriating, internalizing, and ordering multiple layers of social relation.[132] Capitalism operates across complex and interlinked terrains of race, class, and gender in the course of actualizing regional value regimes place to place.

At the risk of accepting the precepts of what historian Donna Haraway dismissed as salvation history—"Can we defuse the bomb in time?"—disalienation must dismantle these multifold hierarchies of oppression and the locale-specific ways they interact with accumulation.[133] Along the way, we must navigate out of capital's expansive reappropriations across productive, social, and symbolic materialisms.[134] That is, out of what sums up to a totalitarianism. Capitalism commodifies everything—Mars exploration here, sleep there, lithium lagoons, ventilator repair, even sustainability itself, and on and on— these many permutations are found well beyond the factory and farm. All the ways nearly everyone everywhere is subjected to the market, which during a time like this is increasingly anthropomorphized by politicians, could not be clearer.[135]

In short, a successful intervention keeping any one of the many pathogens queuing up across the agroeconomic circuit from killing a billion people must walk through the door of a global clash with capital and its local representatives, however much any individual foot soldier of the bourgeoisie, Glen among them, attempts to mitigate the damage. As our research group describes in some of our latest work, agribusiness is at war with public health.[136] And public health is losing.

Should, however, greater humanity win such a generational conflict, we can replug ourselves back into a planetary metabolism that, however differently expressed place to place, reconnects our ecologies and our economies.[137] Such ideals are more than matters of the utopian. In doing so, we converge on immediate solutions. We protect the forest complexity that keeps deadly pathogens from lining up hosts for a straight shot onto the world's travel network.[138] We reintroduce the livestock and crop diversities, and reintegrate animal and crop farming at scales that keep pathogens from ramping up in virulence and geographic extent.[139] We allow our food animals to reproduce on-site, restarting the natural selection that allows immune evolution to track pathogens in real time. Big picture, we stop treating nature and community, so full of all we need to survive, as just another competitor to be run off by the market.

The way out is nothing short of birthing a world—or perhaps more along the lines of returning back to Earth. It will also help solve—sleeves rolled up—many of our most pressing problems. None of us stuck in our living rooms from New York to Beijing, or, worse, mourning our dead, want to go through such an outbreak again. Yes, infectious diseases, for most of human history our greatest source of premature mortality, will remain a threat. But given the bestiary of pathogens now in circulation, the worst spilling over now almost annually, we are likely facing another deadly pandemic in far shorter time than the hundred-year lull since 1918. Can we fundamentally adjust the modes by which we appropriate nature and arrive at more of a truce with these infections?

—*MONTHLY REVIEW,* MARCH 27, 2020

"Internationalism Must Sweep Away Globalization"

Jabardakhal, *the communist paper published in Kolkata, interviewed me on the global economy of COVID-19.*[140]

Jabardakhal (J): *While the British government had initially relied on herd immunity to justify their decision to refrain from preventive measures, the German pro-Nazi AfD (Alternative für Deutschland) activists are demanding authoritarian measures and forced quarantine. The Central Government of India, led by an ultra-right party famous for its authoritarian implementations, has invoked a colonial (1897) "Epidemic Act" in relation to the corona outbreak. The Act includes provisions to give extreme power to bureaucrats, right to detain on suspicion, introduce temporary restrictive and authoritarian measures, and provide legal immunity to government. How do you see such contrasting measures?*

Rob Wallace (RW): Pandemics are mirrors in which countries see themselves. Each ruling party will try to force their country to see itself in the party's premises and platforms, even if those premises, and the actions that follow, worsen the country's outbreak. In the United States, for instance, President Trump closed the border with

Mexico, even as many more confirmed cases of COVID-19, numbering now 360,000, and likely only a tenth of true infections, were already circulating ill-addressed within U.S. borders.[141]

Indeed, for many of these parties, controlling the infection is hardly the first task at hand. Protecting power—or in these parties' minds— *projecting* power comes first. So we have in these mirrors, one after the other, American racism, British Malthusianism, Nazi ghettoizing, and in India's fascist-like Bharatiya Janata Party (BJP), in a Fanonian inversion, a recapitulation of the worst of British colonialism. Indeed, all such sociopolitical manifestations are likely to make pandemic control worse. Neglect, letting COVID run rampant, violating population trust, and jailing people into clusters—each will only amplify the outbreak.

We've seen such outcomes before. As historian Mike Davis describes, 60 percent of deaths from the 1918 influenza pandemic occurred in western India where the British requisitioned food for export during a coincidental drought.[142] At a time in which human solidarity within and between countries is the only path out from underneath a global pandemic, another program in ethnic cleansing pursued in the gruesome Trojan Horse of a deadly virus will only make matters worse. Scapegoating people is no more than a governmental effort to cover up its failure to prepare for the pandemic while pursuing its own Victorian genocide.

J: *There's been a rumor of a biological trade war between the United States and China. It has gained momentum after disputed social media talks on U.S. patents, articles published in* Granma *and mutual accusations of the Chinese Foreign Minister and CIA agents against each other. How do you see this circus?*

RW: Such utterly unfounded accusations are part and parcel of what I call pandemic theater. The efforts we just talked of to control populations within-country are rivaled only by attempts to pin blame for the present pandemic and its socioeconomic ramifications upon other countries. These are all modern updates on calling diseases after an international enemy, now spun into vast, unsubstantiated conspiratorial theories aimed at fast-talking debunkers into exhaustion. What were previously simplified into piquant aliases, such as the Spanish flu or the French disease, are now wound into stories about Wuhan labs or biowarfare gone amok.[143]

Much as for UFOlogy—space saucers, aliens, and the like—perpetrators of such frauds and their dupes are seeking a means by which to avoid grasping the material roots by which capital-led modes of production are increasing our vulnerabilities to the emergence of multiple pathogens of pandemic or near-pandemic capacity. We've seen in rapid succession, upon deforestation and development, H5N1, SARS-1, H1N1 (2009), MERS, H7N9, Ebola Makona, Zika, African swine fever, and now SARS-2 exit out of marginalized wild reservoirs across poultry and livestock and into human populations. Blaming an enemy allows rulers to avoid having to blame themselves for the sudden surge in multiple deadly diseases.

J: *Do you think that the world order is using the coronavirus epidemic crisis to restrict international transactions and promote domestic trade for overcoming the global recession?*

RW: I don't think the world order, if you mean the capitalist ruling class and its state enablers, is in any position to restrict international transactions, however much, for instance, President Trump attempts a nationalist economics the United States no longer has the might to impose upon the world. The virus itself is restricting trade by reducing effective demand, sickening workers, and cutting off supply lines. Hoarding is extending beyond the household to the state with the most ill-prepared nations, among them the United States and Britain, struggling to obtain enough COVID tests and personal protective equipment. Indeed, in the United States, abandoned by the federal government, individual states—New York, California, New Jersey— are competing on the black market for ventilators at many times their actual costs.[144]

The stories are mind-boggling. The comptroller for the state of Illinois raced to meet up with some moving company executive who knew a guy who works with China's factories, handing over a check for US$3.5 million in a McDonald's parking lot, like a drug deal, buying Illinois N95 masks and ventilators, maybe.[145] After being continually outbid by his own federal government for medical supplies and equipment, the Republican governor of Massachusetts hatched a plot with China's consul to the United States and American football team owner Robert Kraft to smuggle in supplies from China aboard the team's private airplane.[146] Jaw-dropping. And markers of a failing nation-state. The richest country in the history of humanity.

J: *Despite the origin of the corona outbreak and a huge population, China has been able to contain the spread of the disease. What is your take on this?*

RW: China bumbled the early days of the outbreak, squashing early whistleblowers, as its instincts tend toward, including a hero doctor who subsequently died from COVID-19 treating his patients. There was a real outside chance that if health authorities moved early enough, the outbreak could have been stanched before it hit the global travel network. No such luck. But upon realization the virus wasn't cooperating with official decrees outlawing viral activity, as if viruses could read, China moved hard upon the outbreak, aiming for total disease suppression: quarantine for infections and case family members, neighborhood blockades, and door-to-door checks. Some of the efforts appeared more actionism than good sense, including keeping people from returning to their homes at night, but not during the day, as if the virus clocks in a day job. But the Chinese state applied the full weight of its resources to match the scale of the disaster. The United States and Britain, on the other hand, refuse such efforts as a matter of realpolitik, having given up on its own public health as a commons decades ago, increasingly neglecting public health or selling it off as a lucrative fictitious commodity.

Why the difference? It isn't merely a matter of neoliberal capitalism as opposed to China's state capitalism. I take the world-systems theorists' position that whereas the United States, as previously Britain, is on the back end of its cycle of capital accumulation, cashing out on public resources, turning capital back into money to be squirreled away in the rich's offshore accounts, China, at the start of its cycle of accumulation, is invested in building its new empire, including the kinds of physical and social infrastructures needed to support such global reach.[147] So while its development, aimed largely at feeding its people domestically, is helping select for the emergence of new pathogens, including COVID-19, China is also invested in suppressing such outbreaks in ways the United States and Britain are utterly incapable of as a matter of structural decay.

J: *Cubans have proposed the use of antiviral medicine interferon alfa-2b as a possible cure. They are training Venezuelan doctors, assisting Chinese medical teams in Italy, and they have even allowed a ship containing corona-infected patients and denied entry by many Caribbean countries to sail to its port as gestures of international solidarity. What is your view on the Cuban outlook of public health, epidemiological research, and global solidarity? What should we adopt from them to combat such pathogenic outbreaks?*

RW: Isn't that quite the inversion? Cuba (and China) sending doctors to NATO member Italy.[148] Senegal turning COVID tests around in four hours, while in the United States, few results are available before a week.[149] Taiwan tests people at the airport for COVID-19, disinfects their suitcases, drives each person separately to your destination in a government-provided taxi, and gives you an app that tells you where in your area you can purchase a mask and another app, slightly creepy, that lists local infections and their case histories.[150] In the States, American passport holders are waved right on through past border officers. U.S. client state South Korea's first COVID case was reported the same day as the United States' first. South Korea's per-capita caseload and deaths—presently 192 total for the entire country—are more than an order less than the 4,000+ deaths in New York City alone. Part of the break with the conservative-liberal consensus against China, New York State's governor accepted 100 ventilators from China.[151] U.S. stature is evaporating in real time. No one is looking to it for assistance or advice.

Whatever its faults, Cuba has long been on the cutting edge of public health innovations despite its comparative poverty.[152] Along with a political philosophy organized around the commons, it excels at the ergonomics in delivering population health services, simple fixes in logistics, and popular education matched with the proper scale of state resources. Not that everything works every time. But it's not in the business of monetizing people's illnesses down to individual insurance plans 28 million people can't afford at all and that another 24 million can afford only in part. So, it isn't merely a matter of what specific drugs Cuba derives. That's terrific, of course. It's the broader ethos that's the matter at hand. A global disease requires a global response. Solidarity is the order of the day. Internationalism must sweep away globalization. And countries that do not partake in such mutual aid will be left to their own devices during their time of need.

—*JABARDAKHAL,* APRIL 8, 2020

The Kill Floor

Everywhere that we examine capitalist regimes, we can see, from macroeconomic policies to the shape and weight of doors and the design of office chairs, a refusal of work mixed in with the overriding obsession with work that is the birthmark of capitalism. In fact, the very logic of capital requires this antinomy. That is why the annals of contemporary capitalism are still written in "letters" of blood and fire and why this will be so until its blessed end.
—George Caffentzis (2013)

Regeneration Midwest, an organization of farmers and food activists across the twelve U.S. Midwest states, held a webinar on COVID-19 and rural America. My presentation, with slides, was dedicated to setting the stage for a broader discussion around what regenerative agriculture could do to alleviate the worst of the outbreak's impacts short-term and long. The text has been both updated from a PReP Rural presentation six weeks later and expanded in its descriptions of the impact on the meatpacking sector.

COVID-19 IS NOW truly a global pandemic.

It's hitting countries Global North and South, although still filling out in South America and Africa, in this, likely, only the first wave of infections.[153] Despite recent declines in media capitals in the United

States and Europe, the outbreak is ramping up at the global level, now clocking in at 100,000 new infections a day.[154] There are already cases of COVID documented even among indigenous groups in deepest Amazonia.[155]

In the United States, some rural counties are now coming in at incidences per 100,000 population that rival New York State, what began as the country's primary epicenter. The rural phase took off this month. We see here on this map the worst time series of reported cases post-May 3, in purple, concentrated in the Midwest.[156]

"The novel coronavirus arrived in an Indiana farm town mid-planting season and took root faster than the fields of seed corn, infecting hundreds and killing dozens," Reis Thebault and Abigail Hauslohner begin their report on rural COVID:

> It tore through a pork processing plant and spread outward in a desolate stretch of the Oklahoma Panhandle. And in Colorado's sparsely populated eastern plains, the virus erupted in a nursing home and a pair of factories, burning through the crowded quarters of immigrant workers and a vulnerable elderly population…
>
> In these areas, where 60 million Americans live, populations are poorer, older and more prone to health problems such as diabetes and obesity than those of urban areas. They include immigrants and the undocumented—the "essential" workers who have kept the country's sprawling food industry running, but who rarely have the luxury of taking time off for illness.[157]

How did this incursion into rural America come about?

To start, as among countries, there is great variation in how states responded to the outbreak. This county map is colored by the date at which stay-at-home orders were announced.[158] The darker the color, the earlier the order. And we can see that several states, including in the Midwest, never issued such orders at all. Sheltering in place is no replacement for mass testing, personal protective equipment, and hospital capacity, but no sheltering assures tens of thousands in demonstrably preventable infections.[159] In the other direction now, some states are reopening more quickly than others. A subset is beginning to lift those orders—for a few economic sectors to start, others the entire state, producing the COVID resurgences models of early reopening predicted.[160]

Why else the variation in local dynamics? Some of the largest meatpacking plants—there in the red dots—are serving as COVID incubators across state, meat type, and company: hog, beef, chicken, JBS, Hormel, Tyson, and Smithfield.[161] According to the Midwest Center for Investigative Reporting, as of June 8, there have been at least 23,500 reported COVID cases directly connected to meatpacking plants in at least 232 plants in 33 states and 86 reported deaths in 37 of those plants.[162]

EWG reports that counties with or near meat-processing facilities host an average COVID-19 infection rate twice the national average.[163] That's a rate that's held steady even when changes in work protocol were implemented after President Trump ordered meat processors reopened in May.[164] From these plants, COVID spreads not only to local communities, but beyond county lines, and by contagion eventually out beyond the commuting range of each plant.

Why meatpacking plants?

While the U.S. outbreak began in big cities on the coasts and was largely driven by the air travel network, the rural outbreak is likely networked by its vast food commodity trade.[165] The two maps here show food tons shipped in 2012 by Freight Analysis Framework area at the top and by individual county on the bottom.[166] So, as Megan Konar and colleagues describe, a shipment of corn might start at a farm in Illinois, travels to a grain elevator in Iowa before heading to a feedlot in Kansas, and then in animal products sent to grocery stores in Chicago.[167] However mechanized the value chain, there are people interacting with each other all along the way. Food commodities are the means by which even the most isolated county can be linked into global epidemiologies.

And we can see the same more specifically for the hog shipments meatpackers tackle. Erin Gorsich's team mapped out a large sample of swine shipments in the United States for 2010 and 2011 based on

Interstate Certificates of Veterinary Inspection for seven major hog states in 2010 with Nebraska added in 2011, together representing 63 percent of U.S. production.[168] The maps here show shipments out, shipments in, and between counties (for the eight states outlined in blue). So, we see large and frequent shipments into and out of these hog centers. At the same time, at the county level, shipment networks are defined by low reciprocity, transitivity, and assortativity. That is, there is a lot of asymmetry in how hog counties are interconnected.

Gorsich's team concludes that the results "are consistent with a vertically integrated domestic swine industry in which large numbers of animals are shipped to the Midwestern states of Iowa, Nebraska, and Minnesota for feeding and slaughter."[169] The researchers conclude that those hubs with the greatest traffic in and out should be targeted for surveillance for such livestock diseases as porcine epidemic diarrhea virus and porcine reproductive and respiratory syndrome. But in a surprise, we suddenly have a virus pinging back and forth this way directly human-to-human.

By what mechanisms is the virus spreading within meatpacking plants? What makes them so infectious? Workers, in effect, are treated as much as sides of beef as the animals they're tasked to process, even, or especially, with an outbreak underway. Workers at the Tyson plant in Black Hawk County, Iowa,

> were still crowded together on the factory floor, in the cafeteria and in the locker room, and most did not wear masks. Tyson said it offered cloth bandannas to workers who asked, but by the time it tried to buy protective gear, supplies were scarce.
>
> At least one employee vomited while working on the production line, and several left the facility with soaring temperatures, according to a worker who spoke on the condition of anonymity for fear of losing his job and local advocates who have spoken with workers at the plant. . . .
>
> On the night of April 12, she said, nearly two dozen Tyson employees were admitted to the emergency room at a hospital, MercyOne. . . .

One worker who died had taken Tylenol before entering the plant to lower her temperature enough to pass the screening, afraid that missing work would mean forgoing a bonus, said a person who knows the worker's family and who spoke on the condition of anonymity to protect their privacy.[170]

There is a political ergonomics to a meatpacking plant that—as reported from Upton Sinclair to Nick Kotz to Ted Genoways—has long pushed off the dangers of the plant and its throughput upon workers as just another price of doing business.[171] Worker safety and health aren't a priority when appearances matter more than the disease itself:

Rafael Benjamin, 64, who worked at Cargill Inc.'s pork and beef processing plant in Hazleton, Pa., told his children on March 27 that a supervisor had instructed him to take off a face mask at work because it was causing unnecessary anxiety among other employees.

On April 4, Benjamin called in sick with a cough and a fever before being taken to the hospital in an ambulance a few days later. He spent his 17th work anniversary at Cargill on a ventilator in the intensive care unit and died on April 19.[172]

Certainly Trump's use of the Defense Production Act to send workers back into these active hotspots is clear as a bell on the point.[173] As are the CDC and OSHA's entirely voluntary COVID guidelines, which now recommend quarantine only when infected line workers are symptomatic.[174] Under the guise of an emergency the administration has also repeatedly framed as "fake news," the USDA allowed poultry plant lines to speed up to rates that require workers to bunch closer together, not farther apart.[175] The Labor Department "all but indemnified companies for exposing workers to COVID-19."[176] The pandemic offers the kind of moment in disaster capitalism that Trump is using to follow through on his promise to dismantle regulations neoliberalism missed.

But it isn't just the White House's fault. States are locking out meat-packers from unemployment insurance to force them back to the plants.[177] Some county health officials, protecting companies, have been documented refusing to share with plant workers how many coworkers have become sick under the guise of "protecting privacy."[178] Other health officials have been frustrated by confusion over which agencies regulate meatpacking operations—the USDA, the state, or local officials—a confusion over which the companies are taking full advantage.[179]

Agribusiness itself is tightening the screws. Tyson reverted to its pre-coronavirus absentee policy, offering short-term disability only after workers get sick with what is increasingly recognized as a chronic condition.[180] The Mountaire Corporation illegally charged poultry line workers for PPE and cancelled its dollar-an-hour hazard pay.[181] "Social distancing," Smithfield Foods' CEO Kenneth Sullivan lobbied Nebraska Governor Pete Ricketts, "is a nicety that makes sense only for people with laptops."[182]

The structure of the industry past and future implies the epidemiological neglect is more than a matter of expedient disregard.[183] Companies warn that plans in motion to automate these plants, an ecomodernist wet dream, are likely to *increase* the price of meat:

The first robot takes X-rays and a CT scan of the carcass, which generate a 3D model of its shape and size. Based on what the system sees in the model, another bot drives rotary knives between the ribs and cuts through the hanging carcass, using the spinal cord as a reference point.[184]

That is, Big Meat admits it has long balanced its profits atop the cheapest labor possible for some of the country's most dangerous work. Peeking ahead, with all those workers no longer necessary, we also infer the next stage in depopulating rural communities.[185]

COVID is meanwhile moving backwards to the front of the supply chain, starting, these early days, to infect large numbers of workers albeit on only a few farms so far, but with three million seasonal workers still crowded together field-to-field and vulnerable to infection.[186]

Elizabeth Royte reports that more than a third of COVID cases in Monterey County in California were diagnosed in farmworkers.[187] At one farm in Tennessee, all 197 migrant employees tested positive, although only three displayed symptoms. In Yakima County in Washington State, 500 cases were reported among agricultural workers.[188] Even when employers respond with handwashing stations, masks, increased social distancing, and temperature checks, the virus continues, as if the problems aren't a matter of individual intervention but systemic, embedded in the business model itself.

The impact of the pandemic was felt at this end of food production long before the virus entered the farm gate. The White House moved to alleviate pressure on the ag sector by proposing to lower the already pittance pay for migrant workers.[189]

Meatpacking and farmwork, whose labor discipline and compensation are holdovers from days of slavery, are treated so shabbily in part because they are dependent upon Black, Latino, and immigrant labor.[190] The demographics are imprinted upon the resulting COVID incidences.[191] Even before COVID, the greater the Latino workforce—particularly in food production and warehouses, now COVID hotspots—the more numerous the injuries logged and the fewer the inspections.[192]

The demographics explain in part why, in a gesture of casual racism, an august Wisconsin chief justice, ruling in favor of reopening, dismissed the risks of the outbreak as failing to threaten "regular folks."[193] At the highest level of governance, Health and Human Services Secretary Alex Azar floated blame for the meatpacking deaths on dirty immigrants bringing COVID *into* the plants, one of a number of novel (and yet very old) contributions to American race science:

> Those infections, he said, were linked more to the "home and social" aspects of workers' lives rather than the conditions inside the facilities, alarming some on the call who interpreted his remarks as faulting workers for the outbreaks. . . .

In a Fox News interview back in April, South Dakota Gov. Kristi Noem explained a massive outbreak at a Smithfield pork-packing facility like this: "We believe that 99 percent of what's going on today wasn't happening inside the facility. It was more at home where these employees were going home and spreading some of the virus because a lot of these folks that work at this plant live in the same community, the same building, sometimes in the same apartment." In an interview with *Buzzfeed*, a Smithfield spokesperson expanded on this theory. Citing the plant's "large immigrant population," she explained that "living circumstances in certain cultures are different than they are with your traditional American family." [194]

Shameless white supremacy in gowns and business suits, offloading blame for the conditions it produced in factory and neighborhood alike. Agribusiness trucked immigrants into these conditions through the H-2A, H-2B, and EB-3 guest worker programs.

In July, Forward Latino and other worker advocacy groups filed a civil rights complaint with the USDA, alleging racial discrimination on the part of meatpacking companies. [195] As Tyson and JBS took millions in federal COVID aid, the companies are subject to federal civil rights law. A CDC report found 87 percent of COVID-19 cases have occurred in minority meatpackers even as they make up 61 percent of the workforce. [196] With the sector's unionism of the business variety, any pushback is laudable. One wonders if such a legal strategy also risks a Dred Scott decision against plant workers in contrast to the bottom-up impacts of the protests against George Floyd's murder that helped inspire the complaint to begin with. As geographers Carrie Freshour and Brian Williams ask, can we better draw on models of resistance from radical traditions openly opposed to the totality of racial capitalism? [197]

Other factors influence rural outcomes. Ag isn't the only source of infection, for one. With the collapse of the town economy, many rural communities turned to another trade in flesh. State and private prisons are major sources of revenue and income. [198] Here the *New*

York Times maps U.S. prison populations and their weekly traffic inbound.[199] We see large prison populations in cities, but clearly some rural counties rival their urban counterparts. Prisons are linked to six of the top twenty-five rural county COVID outbreaks. Close living quarters sicken largely Black and brown inmates transferred across county lines prison-to-prison, as well as prison guards, like meatpackers, moving from prison to community and back.

Comorbidities—other health dangers—and environmental exposures can make COVID outcomes worse.[200] Particulate matter in the air is proving to be one, with an increase of 1 microgram per metercubed in particulate matter of 2.5 micrometers and less associated with an 8 percent increase in the COVID death rate.[201] Christopher Tessum and his group mapped U.S. mortality pre-COVID associated with pollution of $PM_{2.5}$ by locale, economic sector, product, and demographics.[202] We see most of food's contributions emerge out of agriculture—from Iowa, Illinois, the rest of the Midwest, and California's Central Valley—but also commercial cooking and industrial production. Uncharacteristically, compared to other sectors, whites bear the greatest absolute brunt in ag, but Black populations carry a greater load per capita, with Latinos not far behind.

Where are rural COVID patients to go once sick? A failure to access emergency and critical care is also proving a COVID comorbidity. We see many rural counties across the country with hospitals without ICU beds or with no hospitals at all.[203] The North Dakota map shows most counties without a ventilator or no more than a couple.[204]

Big picture, many rural counties represent the bleeding edge of the United States' recent decline in life expectancy.[205] With political consequences. Along the bottom axis of the graph is an index of health metrics around what are called diseases of despair, including obesity, diabetes, heavy drinking, lack of physical exercise, and life expectancy for 2010–2012.[206] The more you go to the left, the worse the index. Up and down on the *y*-axis are the changes in a county's votes for Obama 2012 to Trump 2016. And we see many Midwest states, suffering the worst in public health markers of social abandonment, undertaking the greatest leaps in switching votes from Obama to Trump.

I'm a partisan of neither political party, but clearly economics, health, and the political landscape are fundamentally integrated. We can see these factors interacting together in real time during COVID. In Louisa County in Iowa, south of Cedar Rapids, an outbreak began at the Tyson plant in Columbus Junction, killing two employees, before spreading out to the rest of the county marking Louisa with a per 100,000 infection rate that rivals New York State—with the plant now reopened.[207] Louisa County does not have a hospital, nor a single practicing physician living in the county. As part of a pattern of deception, the state, pursuing a pro-industry public health plan, lied about the seriousness of the initial outbreak at the Tyson plant, reporting 221 employees infected, even as it also had data weeks in hand showing 522 infected.[208] It also reported 444 cases at a Waterloo plant nearby, even as county officials reported more than a thousand.

The Brazil-incorporated JBS hog plant in Worthington, Minnesota, suffered an outbreak that spread out into surrounding Nobles County, now counting nearly 1,600 confirmed cases in a population of 21,629.[209] In the face of evidence that plants were generally more susceptible, JBS kept its line full-tilt through 21,000 hog a day with cases already accumulating among plant workers:

> Van-loads of workers commuted from Sioux Falls [in South Dakota] every day, some of them apparently sick. Employees skipped shifts out of illness or fear. Dozens of workers staged a walkout over lunch to demand that the company slow production lines. The plant's head of human resources abruptly resigned. . . .
>
> In interviews with a dozen meatpacking workers and their spouses last week in Worthington, several themes emerged. They believe the company downplayed the threat of the virus. They say its spread was accelerated by employees commuting from Sioux Falls in JBS-provided transportation. And JBS policies on COVID-19 sick pay and returning to work are applied unevenly, with many fearing that sick people are still going to work.

Maria Echeverria, who works on the second shift and has been at the plant for five years, said managers insisted for days that only one employee was sick with COVID-19, even while it was clear to the workers that more than 30 had it.[210]

Upon closing up in the face of the outbreak, JBS kept Minnesota Department of Labor inspectors out of the plant, a refusal rarely undertaken, suggesting there was something to hide.

The plant may be opening back up to euthanize 200,000 hogs grown too large now without the processing capacity or market to match such a just-in-time supply. "Excess" hog grown out fast and furiously on ractopamine hydrochloride across the country are being gassed, drowned, shot, and, outside the JBS plant, in canonical Minnesota fashion, mulched through a wood chipper.[211] Iowa Select Farms is steaming its hogs alive:

Under [ventilation shutdown], pigs at the company's rural Grundy County facility are being "depopulated," using the industry's jargon, by sealing off all airways to their barns and inserting steam into them, intensifying the heat and humidity inside and leaving them to die overnight. Most pigs—though not all—die after hours of suffering from a combination of being suffocated and roasted to death. The recordings obtained by *The Intercept* include audio of the piercing cries of pigs as they succumb. The recordings also show that some pigs manage to survive the ordeal—but, on the morning after, Iowa Select dispatches armed workers to enter the barn to survey the mound of pig corpses for any lingering signs of life, and then use their bolt guns to extinguish any survivors. . . .

The deployment of armed workers to shoot any pigs who are clinging to life in the morning is designed to ensure 100 percent mortality. But the number of pigs in the barn is so great that standard methods to confirm death, such as pulse-checking, are not performed, making it quite possible that some pigs survived the ventilator shutdown, were not killed by bolt guns, and are therefore buried alive or crushed by the bulldozers that haul away the corpses.[212]

Consolidation, price fixing, and divided supply lines—food service vs. groceries—build in an inflexibility. These hogs (and milk and other food commodities) can't just be redirected to another market or to charity to feed millions of Americans lining up at food pantries in hours-long caravans during the ongoing economic collapse.[213]

So now we can get a better handle on our COVID distribution, shown here just for U.S. rural counties (which may miss the point that urban and rural are in fact integrated systems).[214] In textbook fashion, New York, into which nearly all outbreaks typically get sucked in before being blown back out to the rest of the country, ignited outbreaks in other cities, including Detroit, Chicago, and New Orleans.

But the food commodity circuit also began to spread the virus up and down the value chain, leading to some of the worst outbreaks, culminating in operations in which workers are stacked together as so many sides of beef. The back-to-work order the federal administration is imposing may temporally alleviate the 25 percent drop-off in meat production the food system built in, but only at the expense of more line workers getting sick. Even with sufficient PPE, spaced-apart work stations, and a slowed down line, workers will likely continue to get sick, degrading agribusiness's capacity to deliver.

Will America starve? Probably not, although millions, including in rural counties, go hungry even in the best of times from the very policies Big Ag washes through charity work with food banks.[215] But as with the H5N2 outbreak in 2015, a pathogen is shaking a business model that had long damaged rural areas to its core.[216]

A worse possibility is that the model is under no attack at all. The disease's very danger may have offered the industry a way out of its own trap. We've now learned that the Defense Production Act intervention Trump pursued for meatpacking (but not for PPE production) aimed at protecting Big Hog's access to a bullish China market that lost half its domestic pig supply last year to African swine fever:

Smithfield Foods was the first company to warn in April that the coronavirus pandemic was pushing the United States "perilously close to the edge in terms of our meat supply." Tyson Foods also sounded the alarm, saying that "millions of pounds of meat will disappear" from the nation's supply chain as plants were being forced to close because of outbreaks.

That same month, Smithfield sent China 9,170 tons of pork, one of its highest monthly export totals to that market in the past three years. Tyson exported 1,289 tons of pork to China, the most since January 2017.

In all, a record amount of the pork produced in the United States—129,000 tons—was exported to China in April.[217]

Even with U.S. hog consumption flat since the early 1980s, the sector has recently built out a major expansion—million-square-foot plants and added work shifts—leading to a record 12 percent increase in output 2017–2019. "The producers need exports," the *New York Times* quoted Dennis Smith, a livestock analyst at Archer Financial Services.[218]

We needn't accept any of the nationalist huff implied here to object to the real aim of sending Black, Latino, and immigrant meatpackers into COVID's line of fire: to protect Big Meat's monstrous profits at the expense of the domestic welfare the back-to-work order claimed to protect.

Is there another way? Can what until now has largely been only the hopes and dreams of regenerative agriculture, an alternate mode of production, be turned into a pragmatic transition out of the damage-as-usual? COVID has initiated baby steps, primarily around the sector's logistics.

The Practical Farmers of Iowa have held weekly conference calls on how to run smallholder operations in a time of COVID.[219] Carolyn Betz, one of the researchers for Regeneration Midwest's Health and Climate Solutions research project, attended one of these meetings, reporting back that farmers are working to make both food production and the sales themselves COVID-safe. Farmers are devising modifications to their business practices: for instance, dropbox payments, drop-off Community Supported Agriculture boxes, curbside deliveries, punch cards instead of cash, no-return boxes, and drive-through menu orders.[220]

But how to scale out to match the growing consumer interest in the face of choked off market access? How to escape getting caught in the wreck of industrial production? Some pasture livestock is still beholden to industrial processing. Do we need to purchase a fleet of mobile meat trucks or rebuild regional abattoirs staffed by retrained meatpackers?[221] Do U.S. smallholders need to finally start populating the Open Food Network app that farmers across Australia use to the regenerators' great advantage, including during the present pandemic?[222]

Can we get states and cities to subsidize both the supply and demand ends with price supports for rural farmers and food vouchers for urban consumers, making locally grown foods available to all?[223] Are we ready, finally, to more concretely regionalize supply lines to feed the country in such a way as to prevent diseases in the first place and still feed people when pathogens and other shocks do arise?

—REGENERATION MIDWEST, APRIL 29, 2020
PREP RURAL, JUNE 9, 2020

Square Roots

The civic status of a contradiction, or its status in civic life
—that is the philosophical problem.

—LUDWIG WITTGENSTEIN (1953)

BIOGRAPHER RAY MONK clarified something for me.[224] Philosopher Georg Hegel showed us that different things could also be similar. Ludwig Wittgenstein, in contrast, demonstrated similar things could be different. The square roots of two simple integers, say, 4 and –1, can be very different animals indeed.

How might we apply such a contrast in agriculture?

A coronavirus article I wrote at the start of the year drew greater attention to the increasing capitalization that industrial agriculture and wild foods share, with common trajectories in land use.[225] Weberian farmers meanwhile ping their food animals 'twixt economic signifiers, signaling, back and forth where advantageous, wildness and domestication, including in response to the epidemiological alerts issued upon outbreaks out of either label.[226] There's your Hegel, those two examples.

The Wittgensteinian converse appears demonstrated by industrial and regenerative agricultures. Both produce food, but the first is

organized entirely around alienating social and environmental reproduction and privatizing the resulting surplus value out of their rural
sacrifice zones of origin whatever the damage.

Regenerative agriculture, in contrast, is organized around a natural
economy that supports the capacity of local socioecological systems
to reproduce themselves generation after generation. Use-value
comes first. There are pragmatic advantages too. The approach permits smallholders to trade out exchange for subsistence (buffered by
shared pooling across farms) when the shit hits the fan any one growing season.[227] Or, where Big Ag doesn't choke off market accees, small
farmers might reorganize into passing cooperatives to match a surge
in demand when industrial production can't respond to crises of its
own making, COVID-19 among them.[228]

I would say that at one time the two systems could run in parallel,
whatever their conflicts. One could even be a regenerative capitalist of
a farmer if only by dint of the near limitless bounty of environmental
resources (with intermittent shortages alleviated by jumping to new
land). The latter persona also depended on cheap labor of a variety
of sources: slave, shareholder, migrant, or the self-exploitation of the
family. We mean, of course, *some* societies—the Global North and
their neocolonial satellites—could reproduce this way, externalizing
social damage upon whole peoples even as Earth kept spitting up new
(and often stolen) lands.

Now even that lie is over. With humanity arriving upon a thatch
of global environmental precipices, no truly regenerative agriculture
can be compatible with capitalism. Not even in parallel. The soil,
water, and air that any and all agricultures need are being vacuumed
away at global scales by capital, climate change, pollution, and now
nearly annual pandemic threats. And there's no escape pod off the
planet.[229] Nowhere else to go.

Yet we find both industrial and regenerative ag stuck in a time now
passed. Both pretending they can draw from what they think is best
about the other.

Every agribusiness commercial—already an exercise in alienation cultivating demand—toys with these relationships. Call it

Schrödinger's pig. The ads conflate Big Ag with all the "good" things differentiating such systems from small farming communities—efficiency, modernity, sterility, and the like. At the same time, the ads also claim the very ethos—family farms, the natural economy of sun and seasons—this production destroys.

So we watch widescreen phantasmagorias of economies of scale *and* blue skies *and* "perfect" homogenized produce *and* families of smiling white farmers *and* precision spraying *and* cute piggies held by children. The incoherence is tied off by traumatic bonding between farmer and agribusiness and visceral appeals to a schizoid cultural identity of its own devices.

The ag commercials of this fashion are legion. Indeed, there are no other kind, every one of them mere variations on the theme.

But I find a recent commercial championing mining in Minnesota quintessential.[230] White people saving child slaves in Africa. Basketball legend Kevin McHale. Canoeing pristine waterways. And not a single mine—the ostensible topic—put on camera. An apophatic theology that may speak miner Hades's name but won't show his sulphurous hurl poisoning the state's drinking water.

—FACEBOOK, FEBRUARY 8, 2020

Midvinter-19

SIMON: So we just gonna ignore the bear then?

INGEMAR: It's a bear.

—ARI ASTER (2019)

ASTER'S FOLK HORROR *Midsommar* springs a trap.[231] Its denouement catches audiences on- and off-screen in ambiguous catharsis. When film fans isolated out the swelling soundtrack, the now audible gasps and screams of the men ritually burned alive in that last scene more clearly represent nature's matter-of-fact cruelty.[232] Civilization's reenactments—the movies among them, now shuttered—are unable to keep the brutality at bay. Indeed, with culture as much part of human biology as the enamel on our teeth, the violence, however clothed, is threaded deeper through mind and meat alike.

At present, humanity is caught in its own pandemic theater, with various parties pointing fingers elsewhere to absolve themselves of due responsibility. And we are each prompted to identify with one or another Pavlovian finger. Hero scientists, for instance, are beating back accusations from left to right that SARS-CoV-2, the betacoronavirus behind the COVID-19 infection, emerged out of one of two Wuhan labs rather than in a changing landscape.[233] Which side are you on? In a clickable instant more instinct than ethos, much as *Midsommar*'s protagonist, you've chosen your ingroup.

The ceremonial battles between science and governmental distrust, or, along other axes, China and the United States, Democrats and Republicans, or wildlife vs. livestock, parcel us into manipulable camps. We choose one set of hideous men over another, rubberstamping them all their fight for power over a dwindling planet.

In all likelihood, unbeknownst even to the combatants, evidence is accumulating in favor of *both* hypotheses about SARS-2's origins.

Clearly—to collapse the wave function—the virus ultimately emerged in one, and only one, way. But the growing pair of dossiers, lab and field, should tumble down like an errant curtain to knock all the actors off the stage. A plague on both their houses, some drama queen once wrote.

Until then, against the notion that power depends on knowledge, the two teetering towers of evidence together can only frame the murderous wrestling kayfabe. Real bodies round the world are piling up out of this series of orchestrated clashes.

Did SARS-2 Emerge in the Field?

SARS-2 and the other novel pathogens are not just matters of viruses or their clinical courses.

The webs of ecosystemic relations that capital and state power have pinned back to their own advantage have had a foundational effect on the emergence and evolution of these new strains.[234] Avian and swine influenzas, SARS-1, Ebola, Zika, Q fever, MERS, African swine fever, among many more—the wide variety of pathogens, across different taxa, source hosts, modes of transmission, clinical courses, and epidemiological outcomes—mark different parts and pathways of something of the same regimen in land use and value accumulation.[235]

In China, for instance, agrarian geographer Mindi Schneider finds:

While official party-state discourse conceptualizes "the rural" as a production base for surplus value, and/or as a site for preserving environmental integrity, [my] analysis reveals a further unofficial recasting of the rural: in the process of agroindustrialization, the rural is also a sink for offloading capitalist crises. Between the rivers of manure that flow from industrial livestock operations and contaminate rural waterways; the loss of soil nutrients and food calories in the inefficient conversion of grains and oilseeds into industrial meat; the erosion of agricultural knowledge and practice that accompanies the dispossession of China's farmers; and the shifting values of pigs, pork, and manure, this is a system that "wastes" the rural in service of capital.[236]

We find this new context reproduced region by region the world over.[237] Despite differing in their particularities, local circuits of production operate within the same web of global expropriation and its environmental punishment. At one end of the production circuit, the complexity of primary forest typically bottles up "wild" pathogens. Logging, mining, and intensive plantation agriculture drastically streamline that natural complexity.

Though many pathogens on such "neoliberal frontiers" die off with their host species as a result, a subset of infections that once burned out relatively quickly in the forest, if only by the irregular rate of encountering their typical host species, are now propagating much more widely across susceptible populations.

Ebola strains typically emerge off this frontier of accumulation.[238] Avian and swine influenzas routinely arise at the other end—on factory farms straddling cities.[239]

Other pathogens emerge in more complex origins along these circuits. SARS-1 and now SARS-2 appear to have emerged out of mixed niches spread across their associated regional circuits of production. Nonhuman SARS specimens have been isolated in greater Hubei, Wuhan's province and the first hit by the present outbreak, as far back as 2004, in both bats—Shortridge's horseshoe bat and the greater horseshoe bat—and farmed masked palm civets.[240] The isolates are members of a wide range of animal coronaviruses distributed across

China, including in adjacent provinces Anhui and Jiangxi, well within Wuhan's wild foods catchment.[241]

A trio of Wuhan University management researchers co-authored a paper on food safety in a wild foods commodity chain.[242] The location of the chain is left quietly unidentified, but the discussion is suggestive, identifying six factors influencing disease risk in managing such food chains. Among them:

> At present, the main source of wild-source food is wholesale buyers who purchase directly from farmers or hunters. This method is the main method used to purchase wild-source food in China. In recent years, with the rapid development of cold chain logistics, the source quality management of wild dynamic food has gradually become the weakest and most prone to problems in wild dynamic food express logistics. . . .
>
> Express logistics has high requirements on logistics business and operations, which is often the weak link of enterprises. Proprietary logistics requires huge investment, and outsourcing to professional third-party logistics companies will help reduce production costs, improve management efficiency, and enhance competitiveness. However, the risks of logistics outsourcing are also great.

Clearly, as the paper suggests, the wild food species sold at Wuhan's central market, as elsewhere across China, were raised way out of town before being shipped to the city proper.

But what if it wasn't Wuhan's circuit from which SARS-2 arose? Phylogenetic analyses place SARS-2's proximate origins as far south as Guangdong, the province from which both SARS-1 and several avian influenzas, most infamously H5N1, were originally identified.

Geneticist Peter Forster and colleagues used 160 mostly complete SARS-2 genomes isolated from humans—representing 101 different genotypes across 229 mutations—to produce a phylogeographic network arrayed from China across the world.[243] Such networks permit modeling ancestral and descendent strains co-circulating together, as well as phylogenetic reticulation across samples subjected to horizontal genetic transfer or convergent evolution.

The network parses out SARS-2 into clades A, B, C, with A and B co-circulating early on and C evolving from B. But it's a subcluster in clade A, represented by four Chinese individuals from Guangdong, that hosts sequences closest to the bat SARS outgroup.[244]

The conclusion depends on rooting the bat outgroup the right way, especially as the supplementary material shows those Guangdong sequences were sampled weeks after the first Wuhan ones in clade B.[245] A caveat might be that the bat sample may be able to polarize the character states for network construction here, but may not represent the most recent common ancestor, especially given the genomic recombination events that led to SARS-2 and missing from the bat outgroup sequence. Indeed, new work replaces the 2013 Yunnan bat strain that Forster's team uses with a more recent ancestor, a 2019 Yunnan bat strain from a Malayan horseshoe bat.[246]

From whatever geographic point SARS-2 originated, its genetics—a recombinant of bat and Malayan pangolin strains—indicates the increasingly formalized wild food trade in all likelihood played a foundational role in the emergence of the COVID-19 outbreak.[247] The trade shares with industrial agriculture sources of capital and economic geographies encroaching on Central China's hinterlands.[248]

But how might a bat and a pangolin get together?

Conservation scientist Daniel Challender and colleagues offer a prospective for pangolin farming in China.[249] While the team discounts the likelihood that such farming would displace poaching and trafficking—the main sources of pangolin imports—due in part to care costs and the notorious difficulties in breeding pangolin in captivity, they also report already significant financial investment in China. Even with China's official reserve of pangolin scales supplying traditional medicine, there appears high enough demand, including in affluent dining, to spur taking such farming to scale.

The Guardian reported visiting a farm in Shaoguan in northern Guangdong, previously used for attempting to breed pangolin: "While there were no longer pangolin at the site, several locals near the facility confirmed the species had been raised there, along with monkeys and other wildlife."[250] Other reports indicate such farming operations are only a means by which to launder trafficked pangolin. Chinese conservationists write that previous efforts in pangolin farming were more widespread.[251] Collecting pangolin by any of these modes may offer opportunity enough for such a SARS recombination event.

Indeed, only months before the outbreak, conservation geneticist Ping Liu and colleagues reported discovering multiple bat SARS strains in trafficked Malayan pangolin rescued in Guangdong in March 2019.[252] Upon their rapid deaths and autopsy, many of the captured pangolin expressed the swollen lungs and pulmonary fibrosis associated with SARS infection.

Three of the pangolin strains sampled appeared distally related to a bat SARS collected in 2004 from a greater horseshoe bat in Yichang, Hubei. Another appeared related to a 2011 sample collected from a least horseshoe bat in Shaanxi. A third to that of a wrinkle-lipped free-tailed bat sampled in Yunnan in 2011.

If anything, the diversity of SARS found in this single pangolin batch suggests bat strains widely circulate across provinces—perhaps unsurprisingly for a pathogen on the wing in encroached-upon territory—splattering wildlife and domesticated livestock along the way. Indeed, much as avian influenza has burrowed into poultry flocks as a reservoir of its own beyond the typical wild waterfowl sources, various SARS may soon circulate among food animals, as MERS has among increasingly industrialized camels.[253]

Yunnan bats seem to repeatedly insert themselves in the SARS story, including the ancestral strains for SARS-1, a coincidence that may be a sampling artifact—where researchers happen to sample— but also perhaps because it's true. It is notable that the southwestern province acts as an important nexus through which the Malayan pangolin, whose coronavirus contributed the spike gene to SARS-2, is regularly trafficked:

Zhou et al. (2014) reported on seizures of pangolins made in

China's Yunnan province in the period 2010–2013 (i.e. 2,592 kg scales, 259 whole pangolins). The province of Yunnan shares its border with Myanmar, Lao PDR and Vietnam, and an unknown proportion of the pangolins seized in Yunnan may have been derived from Myanmar or may have passed through Myanmar.[254]

In China proper, a 2016 TRAFFIC brief reports, regional pangolin trade is robust:

Within China, scales were mainly sourced from Guangxi and Yunnan provinces (55% in total), followed by Shaanxi, Guangdong, Guizhou and Hebei provinces. . . .

Besides pangolin scales, researchers also came across one advertisement for pangolin meat (USD180/kg) and two advertisements for pangolin for captive breeding purposes.[255]

The animals aren't just ones and zeros online. "In October 2015," *National Geographic* reports, "customs officials in southern China's Guangdong Province seized a huge haul of pangolins: 414 boxes containing 2,764 (11.5 metric tons) of frozen carcasses."[256] That's a lot of pangolin for a critically endangered taxon.

But perhaps we're missing something. While clearly important to our understanding of the history of this particular pandemic, whether the SARS-2 outbreak began at the infamous Wuhan market itself, farther out on the Wuhan wild foods circuit, or in another provincial circuit altogether, including along the Yunnan-to-Guangdong trade route—all increasingly interconnected by record air and rail travel—is beside the point. Nor does it really matter if bat guano farming, hog husbandry on the hinterlands, or the pangolin trade is to blame.[257]

Instead, we need to readjust our conceptual sights on the *processes* by which increasingly capitalized landscapes turn living organisms into commodities and entire production chains—animal, producer, processor, and retailer—into disease vectors.

Did SARS-2 Emerge in a Lab?

Kristian Andersen and other prominent viral phylogeneticists parsed through SARS-2 gene sequences, coming to the conclusion that the virus's origins were "natural," if one can call the impacts of land use on epizootics we've discussed as such a thing.[258] The virus did not, the team concluded, emerge out of a lab.

Their evidence? SARS-2 is different from SARS-1 and therefore wasn't engineered. The receptor-binding domain in the spike protein that helps SARS-2 enter human cells through the angiotensin-converting enzyme 2 (ACE2) receptor differs from SARS-1's RBD. The team's presumption is that any purposeful engineering would have modeled the virus after SARS-1.

My immediate reaction to the paper upon publication in mid-March was that there is another way by which a lab-specific strain could have arrived upon such a deadly phenotype without modeling itself after SARS-1.

Research and industry labs are moving toward high-throughput screening assays by which multiple drugs and pathogens are tested together in combinations numbering up into the thousands.[259] In effect, science is turning back to nature as an industrial service. Let biochemical cascades run a lab's factorial design at their own kinetics, like a river turns a watermill, albeit aided by automated microplates and artificial enzymes.

Another class of such an industrial model is what is encapsulated as gain-of-function studies that run different strains of pathogen through host species to allow the virus to evolve its own stereochemical solutions to the problems presented by a host immune system or microbicidal.[260]

The grave danger in letting natural selection carry the load is that the labs catching the pathogens on the other end are by definition unprepared for the results. Indeed, the process by which a tested virus might crack a host immune system also permits it a chance to crack a lab's biosafety protocol, especially when such studies aren't always conducted in the best-run operations.

There is the additional problem of so what?

The novel amino acid substitutions recorded in the lab don't capture the totality of possibilities that nature plays with outdoors. Nor do they capture the broader repeated natural experiments that phylogenies of field isolates can. Amino acid residues with the greatest number of coordinated substitutions across genetic loci are also left entirely uncharacterized. Epistatic connectivity is a viral characteristic largely divorced from whatever kind of variation, selection polarity, and conformational exposure catches the typical virologist's eye in the lab.[261]

I wasn't the only one to arrive at the methodological problem with the Andersen team's declaration that no lab accident was possible. Others more recently proposed gain-of-function accidents. Indeed, the notion now stretches from the most conspiratorial takes to the mainstream press.[262]

What I was not prepared for was the pathway by which such experiments were undertaken at the two laboratories in Wuhan, the now not-so-certain city of origin of the SARS-2 outbreak. It wasn't until a week ago that I learned that the EcoHealth Alliance had anything to do with these labs.

EcoHealth Alliance is a New York–based NGO dedicated to deploying principles in "One Health," connecting animal, human, and environmental health to help keep infectious diseases from emerging. In several venues, I've criticized the Alliance and the premises of its work.[263] I was not expecting that the terms of that clash of ideas over the past five years would be reproduced in the current crisis.

The relatively new fields of ecohealth and One Health are modern updates of an old idea that seems surprising only to a society dedicated to alienating itself out of the biosphere.[264] As practiced, however, the concepts as applied by the Global North upon the hot zones of the Global South largely focus on indigenous and smallholder practices at the GPS coordinates at which new infections appear to spill over. Stop eating bushmeat! Stop cutting down forest, local yokels! But such *absolute geographies* miss a critical part of the problem.

Some of us broke off from One Health by way of this omission.

Our take, a more structural One Health, unpacks the *relational geographies* connecting different parts of the world that are driving disease emergence at a much more foundational level of causality.[265] On the global stage, circuits of capital originating out of such centers as New York, London, and Hong Kong finance the deforestation and development driving the emergence of these new diseases at the coordinates that ecohealth investigates.

One can see how an ecohealth or One Health that blames locals for the problem of a disease spillover can serve as a next generation in greenwashing corporate land grabbing. Indeed, EcoHealth Alliance has attracted funding from some of the very multinationals driving deforestation, including Colgate-Palmolive and Johnson & Johnson, two companies dependent upon plantation palm oil. Even now in post–COVID 2020, blaming locals remains a veritable brand for the Alliance.[266]

The interesting epistemological connection here is that such a focus on the forensic particularities of any one outbreak—losing the (sociopolitical) forest for the (local) trees—obfuscates the relations of production that amplify disease spillover. It's only under such a fallacy that a group of ecologists would think running gain-of-function experiments to molecularly characterize SARS to be the way to go.

The SARS-2 escapade didn't start in China, however. We'll pick up the story in 2015 in North Carolina. Epidemiologist Vineet Menachery and colleagues reported on a series of in vitro and in vivo experiments.[267] The team—a collaboration of Chapel Hill and the Wuhan Institute of Virology—placed a series of bat coronavirus spike genes into a backbone of a mouse-adapted SARS:

> In this study, we examine the disease potential for SARS-like CoVs currently circulating in Chinese horseshoe bat populations. Utilizing the SARS-CoV infectious clone, we generated and characterized a chimeric virus expressing the spike of bat coronavirus SHC014 in a mouse-adapted SARS-CoV backbone.

Lo and behold, the artificial constructs proved dangerous:

The results indicate that group 2b viruses encoding the SHC014 spike in a wild-type backbone can efficiently utilize multiple ACE2 receptor orthologs, replicate efficiently in primary human airway cells, and achieve in vitro titers equivalent to epidemic strains of SARS-CoV. Additionally, in vivo experiments demonstrate replication of the chimeric virus in mouse lung with notable pathogenesis. Evaluation of available SARS-based immune-therapeutic and prophylactic modalities revealed poor efficacy; both monoclonal antibody and vaccine approaches failed to neutralize and protect from CoVs utilizing the novel spike protein.

The self-evident nature of the results doesn't match the risk taken. Yes, we learn, SARS can be dangerous to other mammals, including humans. The team insisted on ratcheting up the risk, synthesizing a full-sequence strain based on this combo of backbone and spike, including in live ACE2-competent mice: "Importantly, based on these findings, we synthetically rederived an infectious full-length SHC014 recombinant virus and demonstrate robust viral replication both in vitro and in vivo."

The research group, knowing full well what it risked, speculated that such programs of study were coming to an end in the United States:

While offering preparation against future emerging viruses, this approach must be considered in the context of the US government-mandated pause on gain of function (GOF) studies. Based on previous models of emergence, the creation of chimeric viruses like SHC014-MA15 was not expected to increase pathogenicity. However, while SHC014-MA15 is attenuated relative to parental mouse adapted, equivalent studies examining the wild-type Urbani spike within the MA15 backbone produced no weight loss and replication attenuation. As such, relative to the Urbani Spike-MA15 CoV, SHC014-MA15 constitutes a gain in pathogenesis.

In other words, the team expressed dismay that its dangerous experiment was dangerous, soft-pedaling the implications:

> Based on these findings, review panels may deem similar studies too risky to pursue as increased pathogenicity in mammalian models cannot be excluded. Coupled with restrictions on mouse-adapted strains and monoclonal antibodies generated against escape mutants, research into CoV emergence and therapeutic efficacy may be severely limited moving forward.

What to do then? Back to intonations of a sacred burden of weighing risk and results:

> Together, these data and restrictions represent a crossroads of GOF research concerns; the potential to prepare and mitigate future outbreaks must be weighed against the risk of creating more dangerous pathogens. In developing policies moving forward, it is important to consider the value of the data generated by these studies and if they warrant further study of the inherent risks involved.

The researchers later corrected an omission to the study as to the funding their Wuhan colleague received: "In the version of this article initially published online, the authors omitted to acknowledge a funding source, USAID-EPT-PREDICT funding from EcoHealth Alliance," which went to Zhengli-Li Shi of the Wuhan Institute of Virology.

I believe it a reasonable inference that it was upon the U.S. moratorium that the EcoHealth Alliance, with NIH approval, helped shift funding for these experiments to labs in China, including in Wuhan. To be sure, the Wuhan Institute of Virology has been working on SARS experiments since at least 2008 and the Alliance had been working with scientists in China before 2005.[268] The Alliance had built a long-term relationship with the country's scientific community. Such international collaborations are essential, particularly in studies of pathogens capable of going pandemic.

But such efforts are open to the ugly global intrigue they aim to counter. A follow-up 2017 article EcoHealth Alliance president Peter Daszak co-authored in Wuhan puts the lie to his recent (and unchallenged) claim on *Democracy Now!* that "there was no viral isolate in the [Wuhan] lab":

> Recently we have reported four novel SARSr-CoVs from Chinese horseshoe bats that shared much higher genomic sequence similarity to the epidemic strains, particularly in their S gene, of which two strains (termed WIV1 and WIV16) have been successfully cultured in vitro. . . .
>
> In the current study, we successfully cultured an additional novel SARSr-CoV Rs4874 from a single fecal sample using an optimized protocol and Vero E6 cell.[269]

This set of studies produced constructs that aimed to test the effect of the spike protein on infectivity:

> Using the reverse genetics technique we previously developed for WIV1, we constructed a group of infectious bacterial artificial chromosome (BAC) clones with the backbone of WIV1 and variants of S genes from 8 different bat SARSr-CoVs. Only the infectious clones for Rs4231 and Rs7327 led to cytopathic effects in Vero E6 cells after transfection.

In effect, the team constructed live clones with Yunnan bat SARS backbones and round-robined in spike genes from different bat strains, finding these alternate SARS capable of using the human ACE2.

Upon a 2016 outbreak of bat-related SADS-CoV in Guangdong hog, not far from the locale of the SARS-1 index case, a Wuhan team with some of the same players, including Daszak, conducted an investigation that included exposing piglets to live virus, although the experiments may have been conducted in Guangdong:

> In the first experiment, which was conducted before the virus was isolated, we used three-day old specific pathogen-free (SPF) piglets of the same breeding line, cared for at a SPF facility, fed with colostrum (except one).... The intestinal tissue samples from healthy and diseased animals (intestinal samples excised from euthanized piglets, then ground to make slurry for the inoculum and NGS ... confirmed no other pig pathogens were found in the samples), were used to feed two groups of 5 (control) and 7 (infection) animals, respectively.
>
> For the second experiment, isolated SADS-CoV was used to infect healthy piglets from a farm in Guangdong, which had been free of diarrheal disease for a number of weeks. These piglets were from the same breed as those on SADS-affected farms, to eliminate potential host factor differences and to more accurately reproduce the conditions that occurred during the outbreak in the region.[270]

The worry is that this work led to gain-of-function studies in the more hazardous fashion of Dutch virologist Ron Fouchier, who let lethal avian influenza freely circulate among caged ferrets to self-evolve pathogenicity on its own (in precise opposition to Kristian Andersen's emphasis on forward engineering).[271]

It is presently unclear whether the Wuhan labs—the Wuhan Institute of Virology and the Wuhan Center for Disease Control and Prevention—conducted such off-the-chain studies with SARS strains, for which Daszak and EcoHealth Alliance gave up chain of custody. Indeed, Fouchier himself encouraged just such a follow-up.[272]

I am not convinced that SARS-2 genetics support a lab accident. And one can, and should, reject the Sinophobic weaponization the Trump administration is pursuing, characterizing the EcoHealth Alliance-Wuhan relationship as the nefarious cause of the outbreak and its cover-up.[273] But inquiring after the possibility is still a legitimate endeavor, however loudly scientism rails against theories to that effect, even if it turns out the genetics this round do not support the possibility.[274] After all, the scientific community has long raised the alarm about the biosecurity paradigm in concept and practice.

In 2011, science journalist Laurie Garrett wrote on the "alarming regularity" of accidents in biosafety labs around the world.[275] The accidents are as much a matter of numbers as any one lab's poor safety record. A 2013 Princeton University study showed an increasing global population exposed to the risk of accidents from biosafety laboratories pursuing studies of some of the world's most dangerous diseases.[276]

The study, conducted by health geographer Thomas Van Boeckel and colleagues, showed the population living within the commuting field of BSL-4 labs increased by a factor of four from 1990 to 2012. The fields summed together encapsulate nearly 2 percent of the world's population. Any escape infection could potentially seed an outbreak able to hop upon the global travel network to infect the rest of the world in short order. Since 9/11, *thousands* of new BSL-3 and -4 labs have been built for studying pathogens. The team noted a particular surge in Asia, describing, among others, new labs in Taiwan, Singapore, Pune, Bhopal, and, listed in the appendix, Wuhan.

The lab bubble appears to have arisen out of a combination of legitimate concerns about emergent infectious diseases, the return of "respectable" biowarfare research, the ideological demands of the War on Terror, and the next iteration in Keynesian military economy.[277] The sum effect includes producing the very threat of outbreak the labs were ostensibly set up to stop. Expanding such labs in number and geographic extent bends rare events like viral escape toward inevitability. The labs represent a political logistics folding in calculated dangers that science propaganda spins away as so much conspiracy theory.

In 2004 alone, four separate SARS escapes were reported_out of the Chinese National Institute of Virology in Beijing.[278] In 2018, the Wuhan Institute of Virology used a "less virulent" SARS strain to test lab disinfectants.[279] Are we going to pretend such things don't happen?

You can find a dossier that an anonymous group of self-identified researchers put together in favor of the lab accident theory here.[280] I find some of its inferences and criticism on target and others unconvincing.

Which Option, Then?

What's remarkable is that the EcoHealth Alliance didn't just pay for the collection of SARS samples for Wuhan, the group's putative strength.[281] To circle back to our criticism of its conceptual underpinnings, the organization paid for the gain-of-function studies too. Focusing on the virus or the host or a set of location coordinates or the locals who eat bushmeat or cut down a stand of trees fetishizes causality in the objects involved. Reductionism here, reducing phenomena to the sum of their parts, invades even as relational a science as ecology and health across species.

Meanwhile, on no less than three accounts, a paranoid (or lazily expedient) United States that has now cut off EcoHealth Alliance's SARS grant is in no position to accuse China alone as responsible for the present pandemic.[282]

One, U.S. companies fund or directly own industrial farms gouging into China's hinterlands and, by extension, help drive SARS spillover from bats.[283] Two, NIH money helped back the experiments that the Trump administration now denounces for releasing the pathogen. Three, no innocents they, in 2017, the Trump administration restarted the kinds of gain-of-function experiments stateside it now denounces in China. The relational geographies here extend beyond circuits of capital to include the states servicing them together.[284]

So, what is the Solomonic choice here? All the contending parties—the United States and Chinese governments, the EcoHealth Alliance, scientists, and conspiracy theorists alike—can piss off. Because they told us all to do so ourselves and are attempting to dump responsibility on each other, letting the whole gang escape. Allowing such hubris to continue to run amok would be an act of self-destruction beyond the COVID pandemic and what climate change already has in store for us.

For the dangers extend into the very way we think through these catastrophes. If the system presently organized around global capital—from New York to Beijing—is still positivist, it isn't necessarily empiricist or based on data. Much has been written of China's clampdown on science post-COVID. But such systems of control come in different colors.[285]

Philosopher Alain Badiou writes of liberal parliamentarianism's neo-Kantian opposition of truth and opinion.[286] In the spectacle of debate, we are allowed to struggle over opinion, the core of the extreme center's politics, but truth is never up for grabs, we're told.[287] It's instead locked away in journals and think tanks that the state and philanthrocapitalists own. And the plebs—left coughing on their couches, not from ritual fire but a virus, their pains and sorrows untrue because these are carefully left undocumented—are never allowed access to the official story as a matter of first principle.

Consciously or not, by devious intent or the best of intentions, in accepting what appear diametrically opposed premises that in actuality are scattered together around the dictates of capital and the imperial states that serve it, the various factions of power—money, political class, the latest in colonial medicine—all back some version of expropriation or empire building that regularly spring deadly diseases. By this season, Midvinter 2019, that Great Game finished off a drag race of protopandemic SARS across lab and field alike.

One strain of the many still circulating got out. I'm betting this round that pandemic SARS emerged along the increasingly industrialized wild animal commodity chain from hinterlands and border towns as far south and west as Yunnan. On the last leg of its domestic tour, the virus made its way to Wuhan by truck or plane and then the world.

—PATREON, MAY 2, 2020

Blood Machines

The locals say they're called boilbugs, though no one knew why before now. They breed around streams, carry water inside themselves. Animals eat them during droughts. Usually they're carrion eaters. Harmless. . . . Things change during a Season though.

　　　　　　　　　　　　　　　　—N. K. JEMISIN (2016)

THE BIOMEDICAL INDUSTRY captures half-a-million horseshoe crabs a year.[288] The sector trusses up the living fossils, more arachnid than crustacean, bends their tails back, and bleeds them along a warehouse-length line of steel tables.

The crab's distinctively blue blood, rich in copper instead of iron, is incomparably effective in its immune protection. The defense system's cascade of serine proteases is repackaged as a lucrative commodity for detecting lipopolysaccharide endotoxins produced by *E. coli* and other nasty Gram-negative bacteria. Left untreated on an array of medical devices, including insulin and other injectable drugs, knee replacements, scalpels, and IVs, the toxins can induce sepsis and lead to multi-organ failure and death.[289]

It is astonishing the temporal depths to which we dive for our innovations, in this case, given horseshoe crabs' origins, as far back as 450 million years ago.[290] Our fuels are compressed fossils. Our spaceships are encased in ceramic tiles. Philosopher Alain Badiou sets our agrilogistics as still in the Neolithic.[291] We are at best Aztec runaways in tech and empire, which, while we eat our own ecological tails, is as dementedly aproprioceptive as it is cool.

The horseshoe crab industry, following up exactly on that abandon, is little regulated.[292] After all, the argument goes, the animals are returned to their waters upon what the creatures in some way must conclude is a horror of an abduction. But anywhere from a tenth to a fifth of the animals die from the procedure each year. The survivors, bled a third of their blood for sales of $14,000 a quart, are released with no follow-up on what such a loss—physiological, behavioral, and ecological—means for the population's health.[293] Or for the broader ecosystem, including red knot and other birds that feed on the crab's eggs.

Swiss biotech company Lonza—and now also Hyglos GmbH—developed recombinant Factor C (rFC), a synthetic alternative to what scientists, curling their lips, dismissed as this "backwater" of blood mining.[294] The gene for the Factor C biosensor the crab uses to detect endotoxin in the Limulus amebocyte lysate (LAL) cascade that companies have commercialized to detect Gram-negatives is now inserted all by itself into different kinds of microorganisms for storage and replication. rFC can be deployed as a high-affinity proenzyme in a high-throughput fluorimetric assay. Lots of samples rapidly and accurately tested in a single step.

But U.S. Pharmacopeia (USP), a medical standards group, refused rFC equal standing given its novelty and the crab bleeding system's reliability, even as, in a siloed shortsightedness, the ecology upon which the latter depends may be collapsing out from underneath the medical application. In contrast, rFC has been accepted in Europe.

The clash is finding its next iteration in the fight against COVID-19, as a veritable armada of vaccine and antiviral candidates requires purification, adding to the explosive annual growth in the medical devices sector.[295]

Lonza, Charles River, and Associates of Cape Cod, the three companies that commercialized the crab grab-and-go, say there's plenty of blood supply available. Indeed, they argue, just a day's supply from all three companies would cover five billion vaccine doses. But somewhere the numbers don't add up, with the $1 billion endotoxin test market estimated at 70 million tests a year.[296] Ryan Phelan of conservation nonprofit Revive and Restore balked at the five billion figure on other grounds:

> Ms. Phelan said this calculation "boggles the mind" because, "for every dose going out the door—each manufacturer will use 10 times the amount of LAL to test every step along the way in the process." That includes vials, stoppers and other ingredients in the vaccines. In addition, Ms. Phelan said there are likely to be numerous companies producing vaccines in the test phase and along the way. . . .
> "It is crazy-making that we are going to rely on a wild animal extract during a global pandemic."[297]

Strange to arrive at a place where liberal NGOs, ecomodernists, agroecologists, and Eli Lilly, which has switched to rFC in its COVID antibody testing, might agree: mining horseshoe crabs must end.

The rationales likely differ. The ecomodernists would argue exactly along the lines of the lab meat they approve—fake meat to save real ecologies—except, the agroecologists counter, real meat is needed to save real ecologies, grown in pasture in concert *with* the landscape at concentrations far less than industrial production.[298] The agroecologist would extend the pushback along the lines that fake meat ruins the environment, as it cycles in many of industrial ag's hazardous inputs.

A more probiotic ecology that encircles potential pathogens in complex agrobiodiversities also would better limit disease spillover and virulence. By that higher-order ecohealth, we could reduce demand for our greatly appreciated medical services (which, in turn, could reduce disease load by being made more widely available).[299]

If our story so far seems Jemisin's *Broken Earth* as shot by oceanographer Sylvia Earle, the third act is set by geographer Ruth Wilson Gilmore as a tableau in racial capitalism outside Baltimore.

Photos of the factories show Black labor bleeding the horseshoe crabs row upon row.[300] With modern industry rolling over antebellum labor practices, in the course of producing a clotting agent needed to detect endotoxins and help control a pandemic—both threats that industrial food production itself globally amplified—we arrive upon a biomedical bubble, dumping the resulting damage upon labor on the line, the animals themselves, and local ecosystems.

Capitalism is not just about producing metabolic rifts between our economies and ecologies along the way to profits, destroying our capacity to socially reproduce as a civilization. It's also about producing new ecologies that reproduce capital alienating the web of life.[301] From working people to creatures as old and odd as the dinosaurs.

—JUNE 14, 2020

The Origins of Industrial Agricultural Pathogens

Our group wrote a short chapter for a compendium of work on biosecurity and invasive species still in press. Co-authored with human geographer Alex Liebman, crop and soil scientist David Weisberger, agroecological farmer Tammi Jonas, economic geographer Luke Bergmann, wildlife biologist Richard Kock, and mathematical epidemiologist Rodrick Wallace, we addressed the relationship between industrial agriculture and infectious disease. The expanded version below now includes two additional sections on the historical origins of agriculture, its capitalization, and the subsequent impacts on livestock and crop diseases (and their spillover into human communities).

MODERN AGRICULTURE IS PROVING a leader in the next mass extinction.[302] Along agriculture's advance, primary natural habitat and nonhuman populations are contracting across locales at record rates.[303] At the same time, agriculture is founding new ecologies in their place. Agricultural production and trade promote invasive species and alternate xenospecific relationships, permitting emergent pathogens, pests, and other previously marginalized populations to disrupt long-term ecosystemic function.[304] Indeed, crops and livestock in and of themselves represent the greatest of such introgression.[305]

Succession, one community of species replacing another, is indubitably a pervasive ecological process.[306] All of Earth's biomes originated in part by just such serendipity, with taxa mixing and matching across the globe since life's origins. But there is something foundationally different in the scale and speed at which capital-led production has transformed environment and community alike. In prioritizing commodity profit, an abstraction with foundational real-world impact, industrial agriculture risks undercutting the regenerative capacity of the biological sum upon which even the most human-centric activities, food production included, depend.[307]

So we begin here by unpacking the peculiar nature of such production. We open with a summary of agriculture today, followed by a brief history of how we arrived at such a constellation from cultivation's origins. We next explore the means by which livestock and poultry have been turned into commodities in the flesh. We describe how shifting livestock from the ecology of early husbandry into industrial production helped produce exactly the new barrage of dangerous diseases we now face. We explain how these economies also manifest in crops and their infestation, specifically, in our example, in the emergence of the weed Palmer amaranth.

Along the way, we address how capital monetizes controlling diseases of its own, albeit accidental, making. We propose that however ingenious, the disease control strategies enacted to protect food animals and plants provide nominal defense, acting more as a self-exculpating scientism wielded against alternate food systems. That is, *biosecurity* is an imposition in *biogovernance,* how capital and its allies in the public sector rule societies by intervening into human populations from individual bodies to broader demographics. We argue that biosecurity is deployed first and foremost to protect the most lucrative markets in invasive agriculture.

Invasive Agriculture: Origins and Capitalization

Much research misses that agriculture itself is invasive. The conceptual gap appears to have originated out of a particular cultural context.

Various strains of thought—ancient to modern—have conflated our eminently justifiable sense that we humans have a right to survive with a deeply embedded Prometheanism at odds with environmental conservation.[308] After all, early agriculture—as much an act of desperation as brash innovation when megafauna prey populations collapsed—eventually served as a primary contributor to human expansion.[309] In resetting ecologies in much of their own vision, many a human population began a long demographic payoff in population growth and geographic spread.

Pairing survival and environmental destruction is no necessary connection by any lengths, however. One can accept, indeed celebrate, human survival and reject productivist assumptions about growing food at ever-increasing rates.[310] What foods, after all, can we grow when the inputs we require—topsoil, freshwater, natural pest control, farmer knowledge, among others—have been so degraded by conventional agriculture as to be reconceived as *competitors*, to be replaced by lucrative fictitious commodities such as fertilizers, pesticides, lab meat, precision agriculture, and other technological substitutions?[311] While presently positioned this way in explicit opposition to public health and welfare, food was, and for much of the world's peasantry remains, *agroecological* in origin, tied to the state of the surrounding landscape from which resources are continually drawn (and returned).[312] As a source of organic nourishment that *should* be capped by shelf life, food *should* remain intimately related to biospheric regeneration and local community.[313] In other words, another food world is possible. Indeed, such a pursuit has been long underway in the face of a dominant political economy that for five centuries extended the umbrella of "humanity," on which much environmental discourse still hangs, primarily over the rich and Eurocentric.[314]

These food alternatives do not discount the harsh reality that industrial agriculture—and the interests it represents—presently dominates production to astounding effect. Forty percent of Earth's ice-free surface is dedicated to agriculture, now the planet's largest biome, with many millions more hectares to be brought into production by

2050.[315] Grazing and cropland occupy 24.9 percent and 12.2 percent of global land area, respectively.[316] Poultry and livestock, representing 72 percent of global animal biomass and far surpassing that of wildlife, are simultaneously highly concentrated and widely dispersed.[317] Sixty-four percent of all chickens (22.7 billion global total in 2016), cattle (1.47 billion), sheep (1.17 billion), goats (1 billion), pigs (981 million), and ducks (1.2 million) are found on 2 percent of Earth's land surface.[318] At the same time, 10 percent of these stock are found across 69 percent of land surface.

The resulting impact of such anthropogenic production is geological in scale. While agriculture overall uses nearly 70 percent of global freshwater, the livestock sector uses a third of the water and 50 to 60 percent of cropland and grazing for feed.[319] Feed production, enteric fermentation, manure, animal processing, and transportation produce greenhouse gases at 7.1 gigatonnes CO_2-eq per year or 71 percent of the total added in 2010.[320] Industrial layers and broilers produce 96 percent of global chicken-specific contributions to greenhouse gases (FAO 2019).[321] Discounting intermediate systems, industrial hog account for 54 percent of pig contributions.

How did we arrive at such an agroenvironmental constellation? The Neolithic's early agriculture offered no rival to the scale of present impact. The progenitors launched what turned into a series of reconceptions in ethos and material, one system replacing the next, but, with every iteration in new production still largely constrained by bountiful natural resources, hard to tap and locally bounded.[322] The forces of agricultural production were eventually liberated (even as the relations of production were increasingly, but not completely, privatized into the hands of a select few).[323] Modern agriculture's origins interpenetrated with those of capitalism, the global slave trade, and science, under the decidedly different principle, and double entendre, of grow-or-die.[324]

From the first of Portuguese exploration on, capital-led science was recruited to decode (and recode) the new landscapes and peoples that colonial powers expropriated for commodity production.[325] From the early circuits of capital accumulation across Eurasia and Africa to the Americas, the Caucuses, and the tropics, large-scale croplands and pasture expanded fivefold from 1700 to 2007 to 27 million km², largely replacing forest and savanna/grassland/shrubland, respectively.[326] Capital-led production ramified into regional variations of a globally integrated system.[327] By the primitive accumulation of force of arms and economic compulsion, export extractivism through these circuits intermeshed with a wide array of biogeographies across the diverse combinations of animals and plants that peoples at each locale helped construct on their own terms up to the point of their encounter with capital.[328]

Shifts in food regimes across commodity and geography marked the capitalist period that followed.[329] In the present historical moment, fluctuating centers of agricultural production, networked across global circuits of capital and consumption, source an increasing volume and extent of commodity trade in live animal, produce, germplasm, and processed food.[330] The expansions are tied to monocultures of declining diversity in animal and crop varieties alike.[331] Industrial production of limited lines of monogastric species, primarily pigs and poultry, are replacing locally adapted breeds across nonindustrial countries and into primary forest.[332] Disease ecologist Marius Gilbert's group map the latest shifts from extensive to semi-intensive and intensive production in hog and chicken across transitional economies of the Global South—Bangladesh, Brazil, China, Indonesia, and Vietnam among them—with associated increases in farm consolidation, depeasantization, soil depletion, and manure pollution.[333] Land-use scientist Kees Klein Goldewijk and colleagues project intensively used pasture supporting these livestock at 6 percent of agriculture's land use, converted rangeland 2.4 percent, and unconverted natural rangeland 16.5 percent.[334]

Similar trends are found in crops feeding human populations and livestock. Food systems scientist Luis Lassaletta and his team project feed consumption in hog—tracking a fourfold increase in pork consumption in the past fifty years—will double by 2050 without intervention.[335] Agricultural engineer Ulrich Kreidenweis and colleagues project global cropland area overall will increase 400 million hectares by 2050, mostly in Africa and Latin America.[336] Systems analyst Steffen Fritz and his group mapped IIASA-IFPRI cropland percentage and field size, showing the largest field quartile distributed across croplands in Argentina, Australia, Brazil, Canada, South Africa, the United States, and countries of Eastern Europe and the former Soviet Union.[337] For seven crop-nation combinations, using Landsat satellite data sets 25 years apart, geographers Emma White and David Roy surmise that the increases in median field sizes they report are set by government policy and technological advances, although they refrain from analyzing the capital behind both factors.[338]

Complications abound as field sizes fluctuate place-to-place. Increasing capitalization, for instance, can *decrease* field size, a long-recorded observation, even while also increasing overall land use.[339] Although addressing nothing of ownership, spatial analyst Jie Wang and colleagues report mean rice-farming patch sizes precipitously declining in Jiangsu Province, China, as a result of population pressure, rapid economic development, urbanization, and competition for arable land.[340] Changes in field size, ownership structure, and the diversity of crops planted are often mutually dependent. As with its livestock counterparts, consolidating farmland in industrial countries—reducing farm number while, in the other direction, increasing farm size, input capitalization, and debt-to-equity—has reduced crop diversity between and within species.[341] Of the crops left, cereals—rice, wheat, and maize—have increased in total yield, routinely rolled over in double- and triple-cropping where feasible.[342] Oil crops—oil palm, rapeseed, and soy—have increased in both total yield and harvested area. As a result, industrial agriculture is invading even the last of the equatorial forest.

Under neoliberalism, the latest globalized food regime, the means by which capital grapples with local biogeographies appears to be shifting again. Commoditized assemblages are being reconfigured across borders into multiscalar and spatially discontinuous networks of fluctuating territorial embeddedness.[343] The new multidimensionality is reflected in changes in company management, capitalization, subcontracting, supply chain substitutions, leasing, and transnational land pooling.[344] These state-of-the-art networks are flexibly embedded into non-contiguous biological and political territories, each node only opportunistically attentive to local needs and expectations, although at times agribusiness is forced to be, relying on area infrastructure or a small pool of brand-loyal contract smallholders for local expertise and social brokering. These "Soybean Republics" appear to instantiate the next phase in geographer David Harvey's spatialized model of Marx's general theory of value.[345] Capital production and circulation mold each other, here within a new sociospatial geometry, self-reproducing and spiraling value outward. The resulting core-periphery geographies, at scales domestic and international, are policed along a continuum of violence, from structural impositions felt by individual and population to outright armed repression.[346]

.For the rest of the paper, we will trace the relationship between these new modes of value production in agriculture—livestock and poultry first—and an increasing array of farm and foodborne pathogens. A growing literature suggests that the increasing concentrations of just-in-time monoculture production spread out across the vast commodity chains we discuss here amplify the evolution and spread of the pathogens and pests the agricultures host.[347] We propose these outbreaks represent a second order of invasive species, piggybacking on industrial agriculture itself and spilling over even into local agroecologies that remain only tangentially connected to the capital-led paradigm.[348]

Livestock and Poultry: Husbanding Exchange Value

Sociologist Ryan Gunderson proposes that "industrial livestock production" offers only a *description* of modern protein.[349] Global warming. Carbon and nitrogen budgets out of balance. Polluted waterways. Declining nutrition. New infectious diseases. Morbid obesity and diabetes. Farm consolidation and land dispossession. Rural abandonment. Farmer suicides and an attendant opioid epidemic. The metabolic rifts cutting across these ecosocial domains are only the more morbid *symptoms* of such production.[350] Capitalism, Gunderson argues, is the *explanation*.

Environmental sociologist John Bellamy Foster describes one of the earliest efforts to plot out these connections in what would eventually emerge as critical agrarian studies.[351] Karl Marx originated the notion of shifts in food regime, expounded upon the rifts in capitalled agricultural production, examined the resulting nutrition budgets by occupation, traced the effects of commoditization (and class structure) on oft-poisonous food alteration, and noted intensified husbandry's impacts on proletarianizing farmers, worsening animal welfare, driving out rural populations, and reducing locally available food for the urban underclass.[352]

In their thousands of hectares in livestock monoculture, manure lagoons, impotable waterways, outbreak quarantines, and commercialized rurality, many regional landscapes worldwide today have reproduced (and eclipsed) Marx's map to an alarming degree.[353] In one of the more spectacular examples, China's privately owned Guangxi Yangxiang Co. Ltd is adding to its two seven-story sow breeding operations a thirteen-story "hog hotel" in which to raise a thousand head per floor.[354] Other countries are as invested in such production, if by their own historio-environmental trajectories. Although only the size of the U.S. state of Maryland, with three times the population, the Netherlands effectively internalized the logics of its imperial age, now hosting a livestock sector that generates 9.3 billion euros per year, with production rivaling, and for some species outstripping, far larger European markets *in absolute terms* to grave environmental and epidemiological damage.[355]

Whole counties in the United States are dedicated to the production of industrial food animals at the exclusion of human populations that historically stewarded rural landscapes.[356] As a result, the state of Iowa, a center for livestock and poultry production, is an epicenter for nitrogen, phosphorous, and total solids waste.[357] Its North Raccoon, Floyd, and Little Sioux watersheds, home to 350,000 people total, host the waste-equivalent of Tokyo, New York City, and Mexico City combined.[358] The damage has blown back beyond cutting Iowa's clean rivers in half, polluting private water wells with nitrate and fecal coliform bacteria, and producing nation-leading emissions in fine particulate ammonium nitrate, ammonium sulfate, and hydrogen sulfide.[359] Up against such overhead of its own making, Iowan agribusinesses are now angling to expand their large-scale operations into nearby Minnesota and Wisconsin counties presently characterized by smallholder production.[360]

The effect of industrial agriculture is felt beyond these production ecologies and their sociospatial impacts. Only a subset of Neolithic domestic species able to imprint upon human husbandry proved amenable to being folded further into capitalist modes of production directed at exchange and surplus values.[361] These declining varieties came to objectify—and add to as nonhuman metabolic labor—the fast, cheap, and homogenous growth such a system selected for across a variety of colonized locales.[362] Only a few varieties expressed the morphogenetics flexible enough and reaction norms predictable enough to be developmentally molded for the volume and uniformity of just-in-time markets halfway across the globe.[363]

That is, vast changes have been imposed on *how* food animals are raised (and *what* they even are). Industrial livestock and poultry increasingly incarnate capitalist economy. Genetics, breeding, birthing, finishing, feed, housing, waste management, transport, sacrifice, processing, packaging, and shipment have come to be organized around rates of profit first and foremost.[364] By a modern futures market of nineteenth-century origins, livestock and poultry are now financialized "alive."[365] The animals are committed to fungibilities months before their actual births. These commitments extend beyond their mere identities as commodities on the trading floor. Species are treated as asset classes subject to price volatilities favored as objectively more real than the ecological or epidemiological dynamics into which the actual animals will eventually be born. As a result, breeding, birth, and development are bent to servicing these financial projections first.

Nearly the entirety of the world's poultry breeding by volume is concentrated in the hands of a few multinationals, spurred by a wave of consolidation less about the search for efficiencies than slow market growth and low returns.[366] The primary breeders, who engineer the first three generations of broiler lines that commercial multipliers breed out and market, declined from eleven companies in 1989 to four by 2006: the EW Group, Groupe Grimaud, Hendrix Genetics, and Cobb-Vantress.[367] The ten companies producing layer lines in 1989 were consolidated to two by 2006. The hog sector long trailed its poultry competitors, but by the 1990s entered the Livestock Revolution full-bore. The swine market first surged in multinational breeding companies organized around terminal crossbreeding that maximizes output per animal.[368] A familiar round robin of consolidation followed, leaving a few companies in command, even as farmer cooperatives and national breed societies remain important players: PIC, Smithfield Premium Genetics, Hypor (a subsidiary of Hendrix Genetics), Newsham (Groupe Grimaud), DanBred, and Topigs.[369] The value of the products these primary breeders provide is biologically "locked" by offering the multiplier companies only the males of the male lines and females of the female line.[370] Heterosis-led livestock and poultry—showing the hybrid vigor their parents lack—are treated as trade secrets and must be continuously purchased. By this industrial cascade, a small set of single-source male chicken can generate millions of broiler progeny.

The repeated bouts in market consolidation also helped reduce the sizes and numbers of breeding populations.[371] The sector itself acknowledged the dangers that declines in genetic stock impose upon food availability, epidemiological control, and systems resilience.[372] But company-funded researchers claimed its fiduciary duties, again placing finance first, indemnify the industry of the responsibility of maintaining these genetic resources as a public good.[373] Governments, these researchers argue, must bear the cost of preserving animal germplasm.[374]

Industrial livestock and poultry are bred for fast growth. Poultry thyroids are selected for failing to recognize when poult stomachs are full.[375] As a result, broilers are meeting their much larger target weights—a 400 percent increase since 1957—in only six weeks' time.[376]

The meat is added on so fast that poultry suffer musculoskeletal morbidity and psychopathologies—including tibial dyschondroplasia and stress pecking—associated with growing so much meat on so many thousands of birds bunched together.[377]

Diet offers a means by which agribusiness attempts to close the disembedded livestock economy on its own terms. Industrial ruminants evolved for eating grass are typically fed cheap high-grain, low-roughage diets that destroy rumen microbes, producing ruminal acidosis, a metabolic disorder that would kill the animals save for their subsequent slaughter.[378] To return to the land use with which we began, the shed-led diet explains in part the near-total collapse of cropland pasture in the United States in favor of monocrop grain.[379] No diet is off the table. Since the 1980s, livestock are fed other livestock, helping drive zoonotic outbreaks of neurodegenerative bovine spongiform encephalopathy (BSE) in ruminants and variant Creutzfeldt-Jakob disease in humans.[380] Despite the lessons of such an outcome, and identification of additional BSE forms, there is growing structural pressure to relax the resulting bans in Europe on cross-species and even intraspecies meat and bone meal.[381] Environmental spurs to growth and reproduction extend beyond diet and making feed available at all times. In combination with genetics, counter-seasonal lighting induces out-of-season egg laying, maneuvering layers out of the natural economy of a sun-based photoperiodicity and into continually producing 320-plus eggs during life spans of only 72 weeks.[382]

A veritable pharmacopeia fills in for what genetics and environmental controls miss. For instance, ractopamine hydrochloride, banned in 160 countries, is a beta agonist given to up to 80 percent of U.S. hogs, cattle, and turkey.[383] The drug, originally developed by Eli Lilly as an asthma treatment, mimics stress hormones and relaxes airway muscles. In livestock, it adds more muscle on less feed. Hundreds of thousands of adverse events in the field associated with ractopamine have been documented, including hyperactivity, trembling, broken limbs, stiffness, inability to walk, heat stress and sensitivity, and death.[384]

Livestock and Poultry:
Industrializing Pathogen Evolution

A decidedly monetized husbandry manifests in the epizoologies peculiar to our era. Megafarm conditions select for a wide array of deadly diseases that belie industry's self-absolving invocations of "biosecurity."[385] Among recent emergent and reemergent farm and foodborne pathogens of increasing deadliness and outbreak extent are African swine fever, *Campylobacter*, COVID-19, *Cryptosporidium*, *Cyclospora*, Ebola Reston, *E. coli* O157:H7, foot-and-mouth disease, hepatitis E, *Listeria*, Nipah virus, Q fever, *Salmonella*, *Vibrio*, *Yersinia*, and a variety of novel influenza A variants, including H1N1 (2009), H1N2v, H3N2v, H5N1, H5N2, H5Nx, H6N1, H7N1, H7N3, H7N7, H7N9, and H9N2.[386]

In pitting the industrial engineering of Frederick Taylor against Charles Darwin (and Marx), growing vast monocultures removes immunogenetic firebreaks that in more diverse populations cut off transmission booms.[387] Pathogens routinely evolve around the now commonplace host immune genotypes. Industry densities can also depress immune response.[388] Larger herd sizes and greater densities select for increases in rates of transmission and recurrent infections across pathogen strains.[389] High throughput offers a continually renewed supply of susceptibles at barn, farm, and regional levels, removing the demographic cap on the evolution of pathogen deadliness.[390] Housing such concentrations rewards those strains that can burn through them fastest. As the *persistence* of acute pathogens is also selected for in industrial livestock at population sizes orders of magnitude *less* than in humans, such production can amplify disease spillover into human populations.[391]

Mathematical modeling and a proliferation of data suggest additional epizoological perversities.[392] Decreasing the age at sacrifice—to six weeks in chickens and twenty-two in hog—may select for greater pathogen virulence and viremia able to survive more robust immune systems.[393] "All-in/all-out" production, an attempt to control outbreaks by growing out cohorts in batches, may introduce transmission optima at the level of the barn or farm. The practice may select for a population infection threshold per barn that lines up with finishing times the industry sets for its herds or flocks.[394] That is, successful strains evolve life histories that kill farm animals grown out and near slaughter, when stock are most valuable to farmers. With no reproduction on-site and breeding conducted offshore largely for morphometric traits alone, livestock populations are also unable to evolve resistance to circulating infections.[395] Natural selection is removed as an ecosystem service, no longer able to conduct free work for farmers. The modeling is finding support in the historical record. Increases in avian influenza virulence have been documented nearly exclusively in larger commercial operations.[396] Beyond the farm gate, lengthening the geographic extent of burgeoning live animal trade has expanded the diversity of the genomic segments their pathogens trade, increasing the rate and combinations over which disease agents explore their evolutionary possibilities.[397] The greater the variation in their genetics, the faster pathogen populations evolve.

Even successful gambits in biosecurity carry their costs. By both everyday sanitation and emergency culling, population disease resistance can be filtered out, less virulent serovars that exclude their more virulent competitors can be removed, and exposures that natural immunal development requires can be excluded, including pathogen epitopes that elicit cross-reactive protection.[398] Vaccines—which certainly have their place in food animal health—can also mask or select for the evolution of virulence under industrial conditions.[399] Farms that require stringent biosecurity are typically spatially clustered as a part of the push for economies of scale, producing super-epidemiologies that extend beyond any one farm's capacity to control an outbreak, with decided impacts upon human spillover and pathogen fitness.[400] Some epizoological dynamics are so irregular and unpredictable that the possibility of eradication is eliminated whatever disease control efforts are developed.[401]

Other results signal ongoing disjunctions in scientific thinking. Modeling of Marek's disease virus, an alphaherpesvirus that produces a variety of cancers in poultry, infers that the effects of stocking densities on pathogen mortality may depend on whether layers molt and enter a natural abeyance in laying eggs.[402] Whereas disease modeler Carly Rozins and colleagues argue such a result indicates that improving hen welfare need not be at odds with industry economics, the high stocking densities that led to *reduced* egg loss in the model required the seasonal interruptions molting imposes (and which the layer sector attempts to *circumvent* with counter-seasonal lighting for year-round laying). The researchers' modeling rationale here—searching for room for industrial practices—is particularly loaded given that blaming smallholders for outbreaks and demanding biosecurity protocols that pasture producers are unable to afford, for diseases rarely of their origin, are now part of industry's standard outbreak crisis management package.[403] Development sociologists Paul Foster and Olivier Charnoz and evolutionary epidemiologist Rob Wallace argue such imposition extends beyond a shock doctrine by which outbreaks are used to capital's passing financial advantage.[404] As geographer Alan Ingram and environmental sociologist Marion Dixon describe it, biosecurity offers a mode of governance by which global capital accumulates through nature at smallholders' expense.[405]

The dichotomy between extensive and intensive production upon which industry arguments are based is itself problematic.[406] The essentialist distinction that has been made between industrial farms exercising biosecurity, on the one hand, and small farmers whose herds and flocks are exposed to the epidemiological elements, on the other, is belied by the complexities in ownership and contractual obligations.[407] In many industrial countries, agribusiness ship day-old chicks to be raised piecework by contract farmers. Once grown (and exposed to migratory birds and environmental sources of disease), the flocks are shipped back to the factory for processing. The violation of biosecurity is built directly into the industrial model.

The whole of these constraints—the loss of immune diversity and responsive evolution, the selection for virulence and persistence—are almost entirely self-imposed. Beyond passing off blame, the sector has responded to these disease traps of its own making by modifying food animals. The logistical lengths undertaken in-lab and on-farm, however, are based on expedient readings of the nature of livestock biologies that are also proving gothic in their gore.[408]

"Sterile" hog sheds, for instance, are populated with piglets obtained by "snatch farrowing"—collected directly at birth and reared in isolation.[409] Under the HYPAR or HCDC variations, the piglets are "hysterectomy-procured or -derived" and "artificially reared" or deprived of colostrum.[410] That is, sows that were on the cusp of farrowing are euthanized before or after delivery by terminal hysterectomy.[411] Their uteri are removed and placed in a humidicrib or doused in antiseptic before the piglets are removed from their uterus casing. In some cases, the piglets are latched onto first-generation HYPAR sows so that they can obtain colostrum. Some are medically induced into early weaning. In others, the piglets are fully formula-reared. The aim in breaking the mother-piglet bond is to produce what the industry calls a "minimal-disease" (MD), "high health status," or "specific-pathogen free" (SPF) herd that breaks the vertical transmission of pathogens (and a beneficial microbiome) from sow to progeny.[412] MD-brand pigs are expected to be free of industrially specific brucellosis, enzootic pneumonia, pleuropneumonia, swine dysentery, external parasites such as sarcoptic mange, and internal parasites, including large round worm, nodule worm, whip worm, and stomach worm *Hyostrongylus*.

Post-partum development is defined by another series of surgical and environmental interventions.[413] Piglet tails are routinely docked by cautery iron or cutters, eye teeth removed to stop bored piglets from eating through each other's asses, and castration chemically induced in juveniles.[414] In an effort to conserve farm space and labor, finances again favored first, many industrial sows are placed in small gestation and farrowing crates, in which they are unable to turn.[415] They chew bars out of boredom and develop sores attempting to stand up and lie down. Welfare is almost entirely presented in terms of lesser mortalities for sow and piglet alike rather than animal experience.[416] Industrial livestock are raised more as always-already meat than living animals.[417] The presumption carries over across veterinary care. Subtherapeutic antibiotics in hog, other livestock, and poultry, 80 percent of total U.S. usage in human and nonhuman alike, summing to 34 million pounds a year stateside, are deployed first as growth promoters and prophylaxes for diseases that for the most part would be otherwise preventable by a change in the production model.[418] The agricultural applications help contribute to antibiotic resistance across bacterial infections that contribute to killing 23,000 to 100,000 Americans a year and, with increasing application globally, 700,000 people worldwide.[419]

Such herds, as we noted, are characterized by lesser background immunity. The industrial model of production presumes to codify out the risk of disease by imposing sterile initial conditions in barns full of bodily fluid and manure. The Danish entry/exit is now prescribed: a contaminated area where work boots and clothing are left, leading through an intermediate work area to a clean zone where outbarn wear is provided.[420] Downtime is imposed before visitors are allowed. Showers are provided on-site upon entry and exit. Footbaths are on offer. Production is organized around "all-in/all-out," wherein a herd or flock is brought into a barn only upon the removal of the previous cohort. Barns are outfitted with double-door, "airlock" systems and filters for air circulating in and out. Perimeter fencing and wild bird- and rodent-proofing aim to keep dirty nature at bay. Machinery and tools—trucks, forklifts, ear taggers, and the like—are dedicated by barn. Dead livestock are incinerated, buried, or composted on-site. A separate lunchroom is made available. Processed animal products are banned. Staff are prevented from contact with similar animals— domestic, commercial, and wild—even as sector economics also select for squads of poorly paid itinerant farmhands working farm-to-farm.[421] Interventions extend off-site. Companies have inspected farmworkers' homes as if the source of the problem.[422]

However assiduously followed, even the most stringent programs are proving ineffectual against an increasing array of deadly pathogens evolving, in part as a matter of their own agency. Despite months of warning in advance, wide media coverage, and state and extension program publicity campaigns, highly pathogenic avian influenza A H5N2 burned through turkey and layer operations in the U.S.

Midwest, killing 50 million birds by direct infection or culling.[423] The strain appeared spread by fomites on the wind for which the Danish model offered little protection and the sector's spatial concentration favored. Disease, alongside pollution, farmworker abuse, and all the other rifts we have touched on this article, exemplifies a little-alluded to and intrinsic *diseconomy* of scale. The greater the size of operations at both farm and region levels, the *worse* the problem.

The disease problematic extends beyond the logistical capacity the sector can bear on the outbreaks.[424] In 2016, France's duck and goose sectors were hit across eighteen southwest departments by simultaneous outbreaks of highly pathogenic avian influenzas H5N1, H5N2, H5N3, and H5N9.[425] In response, poultry farmers moved to end industrial production on the basis that the biosecurity practices long pursued were insufficient for biocontrol.[426] In contrast to U.S. production, sector-wide regulation is introducing four-month breaks between flocks during which farmers are expected to clean out and disinfect their barns, a length of time that would effectively end France's place in the global race-to-the-bottom in production ecology.

The quandary is actualized by more than the small likelihood that the usual stable of interventions can succeed in staunching these newly emergent infections. Agribusiness pathogens are also evolving through the core of the model of production as living refutations of a paradigm. The new pathogens, H5Nx and African swine fever among them, circumvent the culturally bound notions of what "biosecurity" must mean to the sector as both a matter of economic necessity and as a "master signifier" on which to ground the story of food for the greater public.[427] These mounds of "sterile" food are proving no such thing. The resulting damage, extending beyond herd and flock loss, has brought an existential anxiety upon industrial agriculture.[428] Dirty diseases that escape putatively clean food "contaminate" industry's narrative during an already ongoing crisis in thick legitimacy and public trust.[429] Fractions of agricultural capital, already clashing over whether to supply highly competitive markets or protect value by planned scarcity, are beginning to lose the kind of class discipline needed to resolve the sector's multiplying predicaments in agribusiness's favor.

Some industrial factions appear adventitious in their response, beyond even the grab-and-go of terminal hysterectomies. Maybe some solution will stick this financial quarter, the thinking goes. Face recognition software for monitoring thousands of livestock for disease or lasers for driving migrating wild waterfowl off farms offer few prospects and involve little in the way of transforming the production schedules at the heart of the business model selecting for pathogen virulence and persistence.[430] "Preprogramming" disease resistance into transgenic food animals aims to block out pathogens before they even arrive.[431] But microbes are much better practitioners of the trick. Across millions of hosts daily, they evolve solutions to prophylaxes many times over before the pharmaceuticals are even introduced.[432]

Agribusiness's capacity to adapt shouldn't be underestimated, however. The sector has long parlayed its intrinsic failure modes in disease control.[433] The ingenuity is astonishing if we miss that the materialism upon which industrial agriculture works extends beyond flesh and machine and into the social and the semiotic.[434] Rather than rethinking the model of production, decades ago intensive husbandry in the United States and other industrial countries spun off growing out animals—what many imagine as farming itself—to contract farmers to bear the worst of disease's losses. Contractors are hired on a gig basis while responsible for millions in U.S. dollars in loans to buy the land, barns, equipment, and other inputs to raise food animals to company specifications. The companies "grow" the contractors, and the capital debts to which these farmers are indentured, to sop up the worst of the damage that outbreaks cause.

Until now the fail-safe has worked as designed. For the H5N2 outbreak in the U.S. Midwest, the direct costs of birds killed by the virus, for which a vaccine produced during the outbreak proved ineffective, fell on the contractors, for whom no outbreak insurance is on offer.[435] The costs of culling flocks as yet uninfected by H5N2 but in danger were paid for by federal taxpayers. In short, to the benefit of the deadliest pathogens, allowed to continue to circulate across a vast network of barns, farms, and national borders, capitalism here, against its own characterization, failed to punish the sector's market failure. The system moves the damage off balance sheets and upon food animals, wildlife, famers, consumers, and communities local and abroad instead. Foundational failures in biosecurity are tucked into production, upon farmers and federal governments first, before a single hog or chicken batch makes it off the truck.

Crops: Palmer Amaranth and Biopolitical Governance

Industrialized crop agriculture offers its own analogs to such spreading zoonosis. In this section, we will examine problems within U.S. commodity crop production related to the expansion of the agricultural weed Palmer amaranth (*Amaranthus palmeri*). This organism and efforts to control it allow us to continue to explore epistemologies around *how* invasion and its management are framed in contemporary agronomic discourse.

Amaranthus palmeri is an annual forb native to northern Mexico and the southwestern United States.[436] It has moved outside of its native range, becoming a weed of concern to major grain- and fiber-producing regions of the Southeast, and has begun a northward expansion.[437] The reasons for this expansion are myriad and center around *A. palmeri's* plasticity and adaptive ability in relation to eco-evolutionary selection pressures, namely changing crop management practices and weather patterns. *A. palmeri* is a dioecious obligate outcrosser—that is, the movement of pollen from distinct male to female plants is needed for fertilization and production of offspring. This facilitates the maintenance of a broad genetic background and is aided by key biological advantages, such as high fecundity—with high levels of pollen and seed production—even relative to other ruderal, primary-successional weed species.[438] These traits allow the weed to maintain high levels of genetic diversity within populations, permitting for the transfer of adaptive traits across geographic spaces that are agnostic with respect to legal boundaries.

Industrial agricultural crop production systems have directly facilitated the evolution and spread of *A. palmeri*. Nowhere has this been truer than the Cotton Belt of the Southeast. In the mid-1990s, the adoption of conservation tillage practices—minimizing plough-ing and subsoiling—and the elimination of more diversified weed management programs, including a reduction in the plant proteins on which the herbicide acted, selected for more robust and competitive biotypes of *A. palmeri*.[439] The reduction in this diversity of herbicide sites of action arose concurrently with the adoption of cotton varieties engineered to tolerate the application of a singular, broad-spectrum herbicide, glyphosate, and was most pronounced in the ten years fol-lowing the roll-out of this technology.[440] The adoption and increase in conservation tillage systems and the use of glyphosate-tolerant cotton were co-facilitating factors, each allowing the other to become more commonplace over the landscape. While goals of reduced soil erosion, carbon oxidation, and energy use were and are valid, conver-sion tillage systems and the reduction of weed management practices, facilitated through the mass planting of glyphosate-tolerant cotton, imposed strong selection pressures on this weed. Undisturbed soil conditions created more advantageous conditions for germination, growth, and development of *A. palmeri,* a species that requires direct sunlight for germination.[441] Seed from this species was now primar-ily left on the soil surface instead of being buried deeper in the soil profile. This alone resulted in larger initial populations of this weed, creating a heightened probability of individuals that carry traits con-ferring herbicide.[442]

Indeed, these traits and individuals were strongly selected for as repeated applications of glyphosate, upon which farmers in the region became heavily reliant, became commonplace both within a given cropping season and across cropping years.[443] Compounding the problem is the reality that weed seeds, many of which now carried traits conferring resistance to one or more herbicides, became mobile entities. While weeds themselves, including *A. palmeri,* are primarily sessile organisms, seeds, and consequently future generations of indi-viduals, including those with traits conferring herbicide resistance, are harvested along with a given commodity crop. This occurs within field boundaries via tractors and harvesters, generating local infestations of various levels of intensity and distribution. Grain or fiber is then transported to both local and distant locations through transnational commodity circulation, turning semi-trailers and rail cars into vectors of dispersal and facilitating the colonization of fields by resistant populations across both proximal and distant geographies.

Much as in the epizootic examples above, weed invasions are not some ancillary feature of such production, but the direct result of the logic of industrial agriculture imposed on the landscape. *A. palmeri* has become a highly problematic weed in the Southeast and has been working its way north, with populations confirmed in areas as far as the Canadian border. In almost every state within the invaded region, both single and multiple herbicide-resistant biotypes of *A. palmeri* have been confirmed, showing that both individual herbicides and mixtures are ceasing to work.[444] That this list is growing suggests a bright future for *A. palmeri*. The fallout associated with this level of human-induced mismanagement and disregard for ecological processes has been catastrophic. Invasions of *A. palmeri* are responsible for devastating effects on agronomic production and economic returns related to major commodity crops such as corn, cotton, peanut, and soybean.[445]

Considerable effort has been spent in describing *A. palmeri*'s development, growth habits, and impact upon cropping systems, as well as in introducing chemical and non-chemical forms of management.[446] Such efforts presently dominate weed management science, extension literature, and popular farm press, the latter two representing major arbiters of information reaching farmers. Primarily, recommendations for the management of *A. palmeri* involve the continued use of herbicides, albeit in increasingly expensive products and mixtures. While such programs may provide temporary respite to farmers, this management paradigm continues to fuel what has become an evolutionary arms race, which leads almost inevitably to more herbicide-resistant *A. palmeri*. Those that stand to gain the most in the long term from this management trajectory are agribusiness giants involved in chemical and biotechnology development and sales. As popular herbicides such as glyphosate have become less efficacious in controlling *A. palmeri* over the last ten-plus years, companies have returned to the promotion and sale of older, more volatile herbicide chemistries (group 4 herbicides, specifically 2,4-D and dicamba), concomitantly breeding transgenic crops modified with tolerance to these chemistries in addition to glyphosate. Increasingly, farmers in the Cotton Belt and elsewhere are moving toward the planting of this iteration of herbicide-resistant crop varieties. This "solution" merely reproduces the same logic that facilitated *A. palmeri*'s exceptional adaptation in the first place. Indeed, *A. palmeri* plants with resistance to these chemistries have already been discovered in Kansas, suggesting a limited timeline for this technology at the very outset of its implementation.[447]

Even more so, the chemical volatility of these types of herbicide active ingredients has led to both production issues and social tensions. Those producers choosing not to adopt these new transgenic crops face the potential for economically destructive crop damage due to herbicide drift across farms. The use of these "newer" technologies has also wrought physical violence between neighboring farmers—invading social norms—and substantially increased reported damage to crops, generating the potential for widespread litigation between industry and individual farmers, as well as between members of the same rural community.[448] In striking similarity to the unevenness in which producers of strawberries or poultry bear the vast percentage of risk in basic production mandates or in adapting to regulatory changes compared to commodity shippers and traders, *A. palmeri* and its management may simultaneously devastate individual farmers and social relations while generating enormous profits for industry through the adoption of new iterations of herbicide-resistant crops and their associated management programs.[449]

Palmer amaranth is commonly understood and managed as an alien phenomenon, a foreign threat to the productive material base of American agriculture, requiring all of the contemporary tools of ecocide and industry. However, the way in which *A. palmeri* is conceptualized, studied, and managed is exemplary of Foucault's notions of biopolitics, the population-level management of plague, and the "abhorrence of contagion."[450] As applied here, such a hermeneutic is informed by recent challenges to the credo of invasive species ecology and conservation biology and recent reformulations of the Foucauldian framework to study the post-9/11 proliferation of biosecurity.[451] Biopolitical governance is explicitly defined as "a matter of organizing circulation, eliminating its dangerous elements, making a division between good and bad circulation, and maximizing the good circulation by diminishing the bad."[452] We can also draw from recent applications of biopolitical frameworks to agrarian landscapes such as California's deregulation of industrial agro-chemical usage conjoined to a devaluation in farmworker health.[453] Here, we reconceive the case of A. palmeri as a specific consequence of mechanisms in biopolitical governance of contemporary industrial grain and fiber production, production mandates, and "invasive species" control, ideologically and materially supported by agronomic science and industry.

The control of weeds and pests at large in highly capitalized industrial agriculture environments reflects the formation of partitions and securitization. The management of A. palmeri—eradication vis-à-vis herbicidal arms race, selective regulation of gene flows, disciplined monoculture—constitutes a regime of governance that enforces utilitarian circulations, among them global crop production, sales of agri-chemicals, and mitigation of capital-suppressing pathogens and weeds, all the while studiously neglecting the ways in which industrial agriculture cultivates the very conditions for these incursions.[454]

There is no lack of empirical clarity as to the underlying causes and consequences of non-cultivated "invaders" in the present industrial agricultural paradigm. Indeed, perspectives from agronomic scientists have stressed the preponderance of herbicide-resistant weeds as a direct consequence of the herbicide industry's impact upon both farm production and the intellectual output of weed scientists.[455] Reframing the coevolutionary interactions between A. palmeri and the logic of commodity production agriculture and biopolitical management offers additional insights into how industrial capitalism fundamentally valorizes different forms of life and evolutionary processes, frequently with disastrous social and ecological results. Political scientist Richard Hindmarsh, among others, theorized the dialectic between containment and uncontrolled gene flow across GM and non-GMO crops as a Foucauldian "escape" from the attempted mechanistic, ordered management of natural resources. *A. palmeri's* escape embodies resistance to the industrial orderings that drove its emergence.[456]

Understanding pest management and agronomic sciences as deeply biopolitical forms of ordering helps to clarify the contradictory directions of regulation and management. On the one hand, crop production helps drive ag-capital's "spatial fixes."[457] By shifts in technology, land price, locational competition, among other factors, a locale may suddenly become transiently conducive to cheap crop production and advantageous exchange.[458] On the other hand, the borders of farm fields are increasingly securitized and surveilled. Furthermore, such a biopolitics depends upon the epistemological disconnect required to create and promote certain invaders while ignoring others, proceeding "not through connections and contagions, but rather [producing] subjectivity through separation and disavowal."[459]

Industrial agriculture exercises biopolitical power through a "more or less stable or shifting network of alliances extended over a shifting terrain of practice and discursively constituted interests."[460] Singular focus on the expansion and optimization of crop populations and export-oriented landscapes—the exercising of the power to "take life and let live"[461]—in the creation of highly capitalized grain agricultures has co-produced the specter of *A. palmeri* and other "invaders." Thus, invasions of *A. palmeri* can be seen most clearly as a Janus-faced mirror, each side reflecting a version of "cultured" and "wild" processes of evolution and ecology, each at odds with the other, up until the damage of one rationalizes the other.

Locating the Outside of a Global Invasion

How might a critical reading of invasion in the context of industrial agriculture and biosecurity help us make sense of emerging invasion phenomena across material, economic, discursive, and relational domains? Upon what do the seemingly disparate H5Nx in France and Palmer amaranth in Iowa converge?

We have argued that invasion is embodied by capitalist agricultural logics and the monocultures, extraction ideologies, and rippling forms of variegated local agricultures tied to global industrial trade. On its face, there seems no "outside."[462] The threat of invasion is simultaneously internal and external to our agricultural systems. It "comes from afar," from futures markets, foreign direct investment, and terminal crossbreeding programs, and is also produced "at home" through intensive monocultures, tillage regimes, and increases in antibiotic, chemical fertilizer, and herbicide application.[463] The constituent components together geologically transform vast swaths of Earth's surface into solar factories, carbon mines, and manure lagoons, an alien landscape hostile to most life forms outside the interest of capital save a subset of suddenly opportunistic pathogen and pest stowaways.[464]

These changes are filtered by locale. Megafarms are sites of mutual construction across livestock, crops, farmer, labor, architecture, and political economy.[465] Their operations effectively divest from the historically ecological and social integration of local landscapes in favor of new programs in biodiversity, water use, outwaste, labor discipline, and economic extraction. Farms are engineered into veritable spaceships, capable—to return to the opening trope—of articulating with any biogeography on capital's terms first (and launching free upon market failure). Earth's ecosystems and local cultures are treated as alien territory, with any resistance to accumulating value to be filtered out. Capital has acted with no compunction in destroying the agroecological and social resilience needed to control regional disease systems before public health or medical intervention.[466] Indeed, as touched on above, nature and all other non-market subjectivities are competitors to be *defeated,* a contrast exceeding even cultural historian H. Bruce Franklin's chilling juxtaposition that under such a system imagining the end of the world is easier than imagining the end of capitalism.[467]

The present program is a losing proposition in the short term as well. The pathogens that arise—H5Nx and Palmer amaranth, among hundreds more that invade the invasion—are integral byproducts of its operations, dependent upon the diverse and proximal entwining with local biogeographies and health ecologies.[468] The feral and the factory interplay atop global capital's peripatetic crest—where a suddenly tightened schedule in poultry production, for instance, selects for the just-in-time ontogeny of avian influenza or where Palmer amaranth evolves virulence or resistance in response to a spike in herbicide.[469] Even without newly moving in, endemic pathogens invade agriculture by evolving out from underneath even the most conscientious biocontrol.[470] "Invasives" also spill back out into the destitute spaces that capital has abandoned by spatial fix, transforming these "ghost" landscapes into new forms.[471]

Invasions also assault the epistemic and the normative, how we think and what we believe.[472] Scientific approaches have selectively overlooked the role of capitalist production—its mechanization, simplification, geographical reordering, and incessant spatiotemporal movement—as a causal factor in the production of invasive species. Indeed, whole classes of modeling, down into their mathematical formalisms, recapitulate such a politics of omission as a matter of first principle.[473] Cost-effectiveness analysis in environmental studies and public health, for instance, aiming to minimize costs in benefits' favor, implicitly accepts the premises of a social system of extreme inequity.[474] Given widespread socioeconomic destitution, the modeling asks, what should we do? The models are organized around an ethics of economism that prudently minimizes expenditures for *some* (less powerful) institutions rather than addressing the structural expropriation that produces the very artificial scarcities the analyses ostensibly target.[475]

Together the scientific, governmental, and industrial responses to pathogenic invasive species constitute new forms of biopolitical governance and new spaces for the expansion of capital—as far down as the molecular level and as ethereally abstract as mathematical proofs. Research is systemically filtered through through the interests of

available extramural funding—the "moneybags"—with agribusiness
increasingly replacing the state as the primary source of funding for
studies in agriculture in capital centers and peripheries alike.[476] While
mainstream scientific literature acknowledges the role globalization
and rising global inequality play in the increasing proliferation and
abundance of invasive species, its conclusions usher in the expansion
of biosecurity logics based in ever-more surveillance and control of
a particular order—ag-gag laws, inspecting farmworker homes, and
microscopic colonialism domestic and abroad.[477]

The new order extends beyond scientific practice at the level of the
investigator. The epidemiological commons we need to protect ourselves from deadly diseases are being sold off as just another series
in proprietary inputs. The rawest of epidemiological data—outbreak
locales, premise identification numbers, shipping records, pathogen
genetics, and transgenic histologies—are being wrapped in a con-
fidentiality that is taking precedence over public health as a basic
domain of intervention.[478] Scientists are increasingly unable to access
even where outbreaks are exactly happening.

Are there alternate programs that can address this triple invasion—
industrial agriculture, pathogens, and commoditized investigation? Is
there an outside to such a global invasion? Is another world possible?
Yes, there is, and it is already underway. Capital, as we described it
here, aims no more than to articulate with the landscapes it arrives
upon as if an otherworldly spaceship. Multinationals and their local
subsidiaries aim to plant and grow out without interference from or
co-optation by local peoples and nonhuman populations alike, save
as a base of resources and labor. The reality is a different matter,
with agroecosystemic players across species coevolving or interacting on their own terms even under the best of capital's applications
in command and control.[479] Other alternatives are more consciously
oppositional, pursuing returns of the agricultural commons.[480]

Agroecological approaches, such as the "push-pull" of semio-
chemical intercropping and trap crops in Africa, diversified cropping
systems, "many little hammers" weed management approaches in the
Midwest, and the farmer-negotiated probiotic ecologies of livestock

landscape health, can foundationally reconfigure agricultural systems in biocontrol's favor.[481] The practices are contingent upon local contexts that refuse to meet global capitalism on anything like its terms as both a matter of principle and practical survival, thwarting the logics of ever-increasing homogenization, scale, and expropriation. Crop diversification, integration of animals into production, and farmer-controlled systems of commercialization and trade also repopulate rural landscapes, undercutting industrial extractivism and biopolitical governance from afar.

While these agroecologies as they are practiced in the Global North have been increasingly cut from their foundationally political bases, there is now a new push at presenting the approaches as an epistemological, ontological, cultural, and political break from capital-led efforts to industrialize natural economies.[482] Agroecology is more than about healthy soil and diverse livestock. Critical evaluations of biopolitics and capitalist geographies need to be folded in as a part of helping reconstitute the community-controlled agricultures that can stem industrial food's infectious diseases.

—JANUARY 20, 2020

Pandemic Research for the People

THE AGROECOLOGY AND RURAL Economics Research Corps (ARERC or "RRC") is announcing a new project.[483]

We're introducing Pandemic Research for the People, or PReP. The project is a crowd-funded effort aimed at immediately getting research efforts underway to answer questions that will help communities around the world during the ongoing COVID-19 pandemic.

There's a lot of great research being conducted right now on COVID-19, from its biomolecular characteristics to potential antivirals to epidemiologies at the broadest geographic scales. But much research remains handcuffed by a political economy going as far back as the origins of capitalism. Funding sources and political appointees gear a lot of otherwise terrific research toward saving the very systems of exploitation that help produce the problems to begin with. Meanwhile, the needs of everyday people most immediately affected, in this case by a pandemic, are left unaddressed.

Pandemic Research for the People will focus on answering questions around the people's pandemic needs first. For this first round, we have six working groups lined up, some with immediate applications, others with the bigger picture in mind:

- *PReP Neighborhoods.* What is the best way to form and safely operate neighborhood brigades checking up on people house-to-house while we all quarantine?
- *PReP Rural.* What can be done to serve the medical needs of rural counties, some of which have no hospitals, ventilators, or even doctors? Can regenerative agriculture help rural communities rebuild local economies at a time when conventional supply lines and end buyers are locked up and more disconnected than ever?
- *PReP Supply Lines.* How might we mass-produce and supply people with enough antivirals, medical equipment, and personal protective equipment to survive the outbreak? What is the broader political economy of global value chains?
- *PReP Outbreak Origins.* How do pathogens such as COVID-19, Ebola, and Zika emerge out of neoliberal land grabs in the first place?
- *PReP Agroecologies.* Could we plan agricultural landscapes that self-regulate in such a way that the deadliest pathogens are far less likely to emerge in the first place?
- *PReP Strategies.* What strategies might we best pursue to break agribusiness and pharmaceutical control of governments and our political economy?

So we have our eye both on immediate needs and the future. Help people now and stop such a pandemic from happening again.

—AGROECOLOGY AND RURAL ECONOMICS
RESEARCH CORPS, MARCH 25, 2020

The Bright Bulbs

*The rest of the world is watching America like
America watched* Tiger King.
—Mr. Onederful (2020)

*I co-authored this roundup of bad COVID takes—right to left—with
agrarian sociologist Max Ajl.*

FROM PRESS TO SOCIAL MEDIA, U.S. president Donald Trump
has been deservedly pummeled for his response to the COVID-19
pandemic.[484] Crackpot denial, diarrheic lying, ad hominem attacks,
gaslighting the masses into accepting mass murder, and not-so-veiled
exhortations to mob violence—a daily skip-to-my-lou through the
DSM-5 and the *Fall of the Roman Empire*. The result? As of this date-
line, the United States has hosted 1.7 million confirmed infections
and 100,000 deaths in four months. One of a new genre of Trump
TikTok imitators nailed the boozy dissolution at the end of the imperial night.[485] "The germ," she tipples, "has gotten so brilliant."

The necropolitics are more than a traffic jam of refrigerated
morgues or, for next month's crop, freedom-lovin' right-wing protest-
ers—"Social distancing is communist" read one sign—congregating

unmasked outside the mansions of Democratic governors and spreading the virus back home.[486] The protest targets themselves, the governors, haven't pursued anywhere near the kind of public health intervention that countries the world over have shown necessary to quash the virus.[487] Public health's abject failure stateside means that any success in wrangling the infection is left largely to millions of freedom-lovin' Americans forced to stay home for many more months than, among other countries, in autocratic China and Vietnam, whose citizens walk freely about.[488]

The point of sheltering in place is to buy time for governments across jurisdictions to stock up and deploy the resources to get the outbreak in hand. Beyond local initiatives, mutual aid campaigns at the neighborhood level, and mad dashes to skirt federal bandits seizing state and hospital medical supplies, little organized follow-through appears evident in the States.[489]

The slapdash efforts offer nowhere near enough testing and hardly any contact tracing. Minnesota, once one of the better U.S. states before bowing to rightist demands into reopening, is only now slowly gearing up for contact tracing, months after the outbreak began.[490] New York City's efforts to hire contact tracers, handed over to the Hospital and Hospitals Corporation instead of the city's more experienced Health Department, appears in shambles.[491]

What are such efforts worth anyway? "We have the best testing in the world," declared the daily-tested Trump about a campaign that ranks per-capita below Belarus.[492] But, in the president's prototypical circumlocution:

Could be that testing is, frankly, overrated. Maybe it is overrated. But whenever they start yelling we want more, we want more. Then we do more and they say we want more. When you test, you have a case. When you test, you find something is wrong with people. If we didn't do any testing we would have very few cases.

The resulting damage, the worst of it entirely avoidable, has been passed onto the labor force's health and welfare. Millions of

Americans, who were unable to afford staying home from work long before the outbreak, filed 40 million unemployment claims in two months.[493] Unlike the wealthy, who have been lavishly subsidized during the outbreak, the United States' "human capital stock" are also denied governmental pandemic assistance.[494] As many as 43 million Americans are projected to lose their job-connected health insurance during the pandemic.[495]

The colossal wreck of it all is especially lost upon crew members. As late as February, the bombastic (and misanthropic) American exceptionalism on parade when COVID-19 first emerged in China set off celebrations in the wardroom, including Commerce Secretary Wilbur Ross's claim the outbreak would "accelerate the return of jobs to North America."[496]

"Most smart people," frequent anti-China Fox guest Gordon Chang tweeted from the communications shack, "knowing that China would dominate the world, thought they should try to manage America's decline. Look who's declining now. It ain't America. Funny what a tiny microbe can do."[497]

Yes, funny that. Even before COVID-19 arrived in the United States, it was apparent that in deploying the virus as a propagandistic parry against China, conservatives and liberals alike would make matters worse by imposing an opportunity cost.[498] By crowding the social space with saber-rattling, the United States would fail to take notes about the outbreak and China's responses pro and con—so as to make adequate and internationally teamed preparations.

Certainly it's a bipartisan miscalculation borne more out of structural decay than mere hubris or bad data, but the problem extends across the sweep of respectable politics. These broader cultural pathologies, entwined into the literal pathologies of the pandemic, are on full digital peacock display beyond Chang and his ilk. The near entirety of the chatter sphere manifests a primal incapacity to adapt a holistic social and ecological perspective. Little subtle, flexible, and capacious thought—capable of encapsulating both the technical-policy-public health sphere of prevention and prophylaxis and the social-ecological-civilizational domain of responding to the problem from the bottom up—is on offer. No one is caught dead taking the lead of the world's most affected dispossessed, who might know something about such thinking.

One instead traces a long arc of incompetence, reductionism, social triage, capitalist Mad Hatter logic, technicist tomfoolery, and rank opportunism. From the privatized right to the public left, influencers political and academic have been studiously incapable of responding to the crisis. Against all notions of political ecology, the pandemic is a "Chinese virus" on the right wing or an "act of God" on the left, removing off the board any notion of refounding our agrarian practices or the other modes of social reproduction that together drove the emergence of COVID-19.[499]

Dinner Party Politics

Among liberals, Chang's misaimed schadenfreude is directed through more learned lenses at the hoi-polloi.

Max Fisher wrote in the *New York Times* mid-February of a dinner that two professors at the University of Washington held for students and faculty, including one student who railed against the panic over a virus that had at that point killed 1,100 people worldwide, while influenza kills 400,000 a year:

> Ann Bostrom, the dinner's public policy co-host, laughed when she recounted the evening. The student was right about the viruses, but not about people, said Dr. Bostrom, who is an expert on the psychology of how humans evaluate risk.
>
> While the metrics of public health might put the flu alongside or even ahead of the new coronavirus for sheer deadliness, she said, the mind has its own ways of measuring danger. And the new coronavirus disease, named COVID-19, hits nearly every cognitive trigger we have.[500]

There were, of course, no consequences to Professor Bostrom's incapacity to gauge Professor Bostrom's incapacity to evaluate risk.

She has since been repeatedly quoted by *Science* magazine, now berating everyday people in the other direction:

> It will take more than just messages to change behaviors on such a mammoth scale, says Ann Bostrom, who studies risk perception and communication at the University of Washington, Seattle. Often, compliance hinges on giving people the tools they need to easily follow new rules. "The physical context in which you make these decisions is often more important than grand ideological views," Bostrom says.[501]

Indeed, while Dr. Trump's bleach and light therapy cures sent double the usual poison cases to the hospital in New York alone that week, the outrage also offered Democrats and American public health more broadly the kind of cover they needed.[502] In pivoting to White House quackery, the self-dubbed "resistance" could avoid its learned impotence in reversing an outbreak in part its own making. The provocation serves all parties in the pro wrestling storyline. As regularly reported on since 2016, wrasslers of both parties will meet up later for cocktails—what's your poison?—in a reopened Hamptons.[503]

The clash is one largely over style and emphasis. Under bourgeois political economy, the state, steered by either party, is expected to act as capital's handmaiden. The Democratic critique orbits around the notion that the present administration is little versed in these duties. Outside doling out government checks to sector after cursed industrial sector, a holdover from the Trumps' days of bribing tax assessors in New York, a family that has had every piece of underwear picked up by staff has proven itself unable even to envision what's involved in servicing the logistics and infrastructure that capital requires to accumulate from one side of the world to the other.[504]

The Trumps abandoned what every cabinet acted upon since the Second World War. Global capitalism is an American project and the United States is supposed to keep the wheels greased for the bourgeoisie regardless of nation of origin.[505] The world is, or suddenly only once was, the U.S.'s client state. In fighting other states over petty bilateral trade, the United States has lost the begrudged loyalty it owned nearly lock, stock, and barrel. American power, for instance, is on the hook for cleaning up pandemics that capital the world over helps create in order to keep the system on the same developmental path despite the ultimately catastrophic alienation in land and labor.[506]

Under Donald Trump, the United States has abandoned controlling these outbreaks, even one so deadly now on its shores. As he has done repeatedly in his professional life, Trump filed his administration's failure of a COVID response for bankruptcy. After first floating, then retracting, Jared Kushner and Ivanka Trump's names for a council to reopen America out of an uneven quarantine, the president read an exhaustive list of corporate executives and companies to staff the council, many of them surprised to be on the list.[507] The council's objective? Close down the coronavirus task force—an option Trump reversed upon widespread complaint—reopen beaches, barber shops, and meatpacking plants; roll back infection control in nursing homes; and let someone else pick up the toe tag of a bill.[508] With the virus marching on, the administration aims to cut the losses in pandemic PR and double-time millions back to work.

Liberals rolled their eyes.[509] In comparison, with the characteristic panache of its commander-in-chief, Obama's administration expertly delivered cover to agribusiness for swine flu H1N1 in 2009 and multinational development in Guinea, the country of origin of the Ebola outbreak in West Africa.[510] Indeed, these yesteryear triumphs are embodied by presumptive nominee Joe Biden, who, repeatedly citing his own verve in handling Ebola as vice president, is now running to the right of Trump on China and is himself in the get-back-to-work camp, although with public declarations about getting the outbreak under control.[511]

Marching the country back to work almost certainly was the substance of Biden's call with Trump in early April.[512] Despite the contrary characterizations of the chat—Trump got Biden's approval or Biden told Trump what to do—the phone call effectively offered the political class the imprimatur it needed for sacrificing hundreds of thousands it refused to protect here in the United States, much less the world. The publicly announced back channel aimed to assure the bourgeoisie their interests would be served across party lines.

Fauci's Donuts

Establishment science, repeatedly positioned as anti-Trumpism incarnate, is also stumbling about imperial epidemiology's biggest stage.

When Ma Xiaowei, the head of China's National Health Commission, announced in January that SARS-CoV-2, the virus that causes COVID-19, could be transmitted before symptoms appeared, U.S. epidemiologists attacked.[513] They denied the possibility and demanded access to the data that led to such a conclusion. China has not accepted U.S. offers of research assistance.

One-time Oprah guest and epidemiologist Michael Osterholm weighed in this way:

> "I seriously doubt that the Chinese public officials have any data supporting this statement," said Michael Osterholm, director of the Center for Infectious Disease Research and Policy at the University of Minnesota. "I know of no evidence in 17 years of working with coronaviruses—SARS and MERS—where anyone has been found to be infectious during their incubation period."

Two days before, a research team from China published just such evidence in *The Lancet*.[514]

Grappling with a novel pathogen is dicey business. Everyone gets surprised along the way. But the failures to respond well are often buried deep in a system's political epistemology that is no better personified than by NIAID director Anthony Fauci, liberal hero whose mien, already played by Brad Pitt, graced donuts across the eastern seaboard.[515]

As he has survived multiple administrations—from Reagan to Trump—Fauci is inherently a political animal. On its face, that seems a desirable attribute for someone charged with protecting technicist intervention against administrative interference. We need ask instead what polite refutations mean under an administration that tailors both public health messaging and policy to the political needs of the Oval Office alone. Miming counter-governance without power is exactly the kind of "resistance" that has only repeatedly underscored Trump's albeit dysfunctional control the past three and a half years. Fauci received considerable praise for contradicting administration talking points before the House Committee on Oversight and Government Reform, but exchanges like this are little but astonishing:

REP. GERRY CONNOLLY: Was it a mistake, Dr. Fauci, do you believe, to dismantle the office within the National Security Council charged within global health and security?
DR. FAUCI: I wouldn't necessarily characterize it as a mistake. I would say we worked very well with that office. It would be nice if the office was still there.[516]

Even sympathetic Democrats lost patience with the charade. Despite bumping elbows with the director later, during testimony, Congressman Stephen Lynch (D-Boston) admonished:

Dr. Fauci, you've been great on some of this stuff and pushing back when the president said, "We're going to get a vaccine fairly quickly, a matter of months." You were good to step up and say, "No, it's going to be a year and a half." But we really need honesty here. And when the president is making statements like this, we need push-back from the public health officials standing behind him and nodding silently. An eye roll once in a while is not going to get it.[517]

No one need dismiss the difficult spot Fauci finds himself, trying to steer a public health response captained by a meglomaniacal dunce. And maybe that exchange with Lynch was a setup for giving Fauci more political room. But he has helped steer that ship, straight into an iceberg.

All the old tricks aren't working, if they ever had. The nation's obscene response to the HIV epidemic served as Fauci's first portfolio. Medical humanities scholar Graham John Matthews reminds us:

Early epidemiological data from Africa were wrongly interpreted as "proof that 'household contact causes AIDS.'" Anthony Fauci, in his editorial for the May 6, 1983 issue of *JAMA*, reinforced this hypothesis by citing the possibility that "routine close contact, as within a family household, can spread the disease. . . . If we add to this the possibility that nonsexual, non-bloodborne transmission is possible, the scope of the syndrome may be enormous."[518]

Bad medicine is routinely tied to bad politics. "Dr. Fauci, and the medical and research establishment," political scientist J. Ricky Price writes, "rarely address systemic structural issues of poverty, racism, misogyny, and transphobia as concurrent problems in the fight against AIDS. Nor do they mention the predominant strategy that the U.S. legislative and judicial branches have used for prevention: criminalization."[519]

In another recent appraisal, sociologist Trevor Hoppe found a familiar playbook for the Ebola outbreak that began in West Africa in 2013:

> Maine governor Paul LePage threatened to take action but hesitated to follow New Jersey's lead in instituting mandatory twenty-one-day quarantine policies for anyone who had been in contact with Ebola patients after Centers for Disease Control and Prevention (CDC) director Anthony Fauci called such policies "a little bit draconian."[520]

Spinning scientific outcomes into diplomatic service appears a reflex. In a commentary published just two weeks before COVID-19 exploded in the United States in March, Fauci and his co-authors bright-sided two recent reports on clinical cases:

> If one assumes that the number of asymptomatic or minimally symptomatic cases is several times as high as the number of reported cases, the case fatality rate may be considerably less

than 1%. This suggests that the overall clinical consequences of COVID-19 may ultimately be more akin to those of a severe seasonal influenza (which has a case fatality rate of approximately 0.1%) or a pandemic influenza (similar to those in 1957 and 1968) rather than a disease similar to SARS or MERS, which have had case fatality rates of 9 to 10% and 36%, respectively.[521]

The commentary massaged COVID's case fatality ratio down to influenza, again, to no professional consequence, fattening donuts notwithstanding. It left out what no vaccine and no herd immunity mean for such a number. That is, even under presumptions of an R-naught and a CFR comparable to influenza, it was understood early on that the pathogen could still exert considerable damage.

And why set COVID against influenza, the rookie mistake comparing diseases measured by different methods and at different points of their epicurves, as if they also aren't killing people side-by-side?[522] Why not discuss the accelerating emergence of multiple spillover threats?[523] African swine fever, yellow fever, H7N9, and, among many others, Ebola, just ending yet another run of *years* in the Congo following the West Africa outbreak?[524] Why apply biomedical epistemology to disease ecology and political economy? Why presume centroid measures of CFR also mark its statistical spread? Why presume such dispersion captures the amplitude (given the virus's global penetrance)? Why presume the sample variance Fauci's group comments on reflects the population dispersion (or even that of other samples)?

Why presume future trajectories under such an evolving distribution will be like those of the past? Why presume said trajectories mark some set taxonomic character on the part of an evolving RNA virus, as Osterholm did, with the still-debated possibility of reinfection and long-term infection and damage, turning SARS-2 into a chronic disease?[525] Why place causality largely in the object of the virus or infection and not in the field of epidemiological opportunity, where, as the commentary does say, we can help set things better?

Why does such bright-siding, delivering good news first, presume operating under the risk of a Type I error—failing to prepare for a true danger—makes for better public health strategy than under a Type II risk—seriously preparing for an outbreak that turns out a nothingburger? In addition, against scientism, why presume data at the level of a clinical study are enough to inform policy?

Why not err on the side of precaution? As with the 1976 influenza outbreak, Type IIers might have been laughed at if COVID proved little danger, but at least researchers of this persuasion took people's lives seriously enough to champion preparing for the worst from the start (even as they might also hope for the best).[526] Such decisions have to be made early on. Unlike what Trump claimed in February to questions about dismantling the U.S. pandemic response and firing all those epidemiologists since taking office, a country can't just order up a public health response like pizza to go should the outbreak turn uglier than it first seems.[527]

With the COVID-19 pandemic long escaped and the U.S. months behind in response, if ever to catch up, Fauci appears the epidemiological equivalent of another bipartisan stalwart and too late convert—Alan Greenspan during the housing crisis.[528] Perhaps Fauci only now is beginning to grasp that the premises of a political metaphysics that brought us to this moment do not apply to its resolution.

Such a compensatory fantasy may be giving the director too much credit. Fauci has repeatedly claimed he's never been muzzled by the president, apparently muzzling himself. Commentator Jeffrey St. Clair got at the crux of the matter:

> If what Dr. Anthony Fauci said on Monday was true (i.e., that Trump immediately followed every one of his suggestions), then he should resign for failing at his job. If Fauci lied, then he should resign for allowing himself to be infected by Trump's own pathological virus.[529]

One suspects both cases likely. Corruption isn't just a matter of envelopes passed under the table.

That happens too, of course. Globally renowned epidemiologist John Ioannidis led a Stanford University study modeling COVID-19 that was published as a preprint in mid-April.[530] Like the Fauci commentary, the study downplayed COVID mortality rates as little worse than flu.

A whistleblower complaint to the university showed a round robin of emails among Ioannidis's team to the effect that the study operationalized its parameters on the basis of several thousand antibody tests conducted in Santa Clara, California, using a kit Stanford University microbiologists concluded erred on the side of false positives.[531] Alongside the obvious impacts on the model parameters, for one, increasing the case fatality ratio's denominator, study scientists also complained of the impact of reporting back unsubstantiated test results to study participants.

But perhaps worse was the political economy in play. The complaint also showed the study was funded in part by and conducted in coordination with David Neeleman, the JetBlue Airways tycoon and reopen proponent. Despite a bailout of astounding proportions, the airline sector as it is presently constituted is on the verge of collapse.[532]

Red Vegans Against Green Peasants

The bad takes on COVID wend across the political continuum into the more recognizable left. Superficially more grounded anatomies of the crisis have leapt atop the backs of the dead animals and broken landscapes that did indeed help produce the pandemic. But in a classic riding trick, the acrobatics suddenly switches mounts mid-ride to characteristically Eurocentric hobbyhorses from which to herald imperium-old edicts on how to live, eat, and die.

Should we eat meat, with source livestock an apparent driver in the emergence of deadly pathogens? Documentarian Astra Taylor, environmental historian Troy Vettese, and political scientist Jan Dutkiewicz—TVD, for brevity's sake—answer in the negative: "Individually, we must stop eating animal products. Collectively, we must transform the global food system and work toward ending animal agriculture and rewilding much of the world."[533] With anthropogenic global warming already taking carbon dioxide levels through the roof, meat was already an easy target. It's a synecdoche for effete gluttony, the emblem of a global class divide, an easy piece of fat—and protein—to trim from wealthier consumption baskets, and a neat way to merge individual ethical consumption and world ecology.

The anti-meat crusade has apparently received an unintended and misused push from recent ecological and epidemiological work on the likely origins of the pandemic.[534] These analyses traced how the interaction across confined animal feeding operations, monoculture doppelgangers, fading forests, and antimicrobial marination has produced a petri dish of new diseases. Out of this combination, one virus after another easily jumps from animal populations to humans.

Pre-pandemic, TVD fellow travelers rejected such political ecology, which in their psychologizing dismissal "often romanticizes what are seen as anti-modern subsistence livelihoods on the margins of global capitalism."[535] But now, given the obvious realities on the ground, a pandemic strain that hopscotched from bat caves on the other side of the world into the lungs of urban workers they champion, the ecomodernists (again to no reputational damage given their golden tickets) have turned to folding in the analyses they previously characterized in the most scurrilous terms as if they approved all along.

Such systems—these incubators for viruses, huge biological emitters of CO_2 and methane, rampant deforesters, and living beings suffering amid the cruelty of enclosed industrial animal camps— merge into a pithy command from TVD: No, don't eat meat. The team suggests we plow "public-directed investment" into "both plant-based meat alternatives and cellular agriculture," or, in other words, lab meat, a product that so far exists primarily among venture capitalists, a few labs, and red-washed ad copy lauding it as a socialist wonder food from Keynesian Green New Deal cookshops.[536]

Key questions are greased over, restricting, as sociologist Andy Murray describes, the very discourse lab meat proponents claim they wish to open up.[537] Who is this "we," for one, and even, what is meat? Veganism and animal rights, to which one needn't object as ethoses on their face, are reflexively deployed here to conflate objects and processes. There is no *thing*, meat, that has uniformly negative ecological, social, or epidemiological consequences. Meat only has in common that it comes from living creatures, and animals, just like people, can only be fundamentally understood in relation to the material environments within which they live, are loved and cared for, or maltreated and abused, and, in the case of most food animals, killed.

The question of "Should we eat meat?," therefore, appears very different among different sets of "we" and the different relations "we" have with such animals.

There are millions who might bridle at, or whose lives would be simply upturned and devastated by, enforceable commands that they simply cease meat production and consumption. Tunisian camel herders in the semi-arid steppes of the Jerid who rely on herding for day-to-day survival, or Bedouins in the northern Gaza Strip, have not been consulted about how they feel about an order from the Global North—in this case from Harvard and Johns Hopkins direct—to stop eating meat or engaging in the meat trade.[538] Nor, in the other direction, have these researchers asked if such meat is substantively identical to the confined feedlots they rightly condemn.

At a minimum, we know that ceasing meat production and consumption would require a massive political intervention in those countries. We know that isn't what the authors intend, heaven forfend. But we also know that intention does not get us very far, particularly when the *Guardian*, where the TVD piece was published, has rarely shied from advocating for neocolonial assaults on the Third World.

Beyond that coliseum, it would not be the first time that phantasmagorical narratives of "environmental degradation," "resurrecting the granary of Rome," or "making the desert bloom" have been used to justify the extirpation and violation of the rights of Arab people in the peripheries of the world system.[539] The impulse is little different from the One Health approach, which, connecting wildlife, livestock, and human health, warmly speaks of "the creation of a healthy and sustainable reconnected future for our planet."[540] Indeed, many a leftist outlet—Sonia Shah interviewed on *Democracy Now!*, for instance— platforms such cant in COVID's wake.[541] In actuality, the approach aims to recapitulate colonial medicine, blaming local indigenous and smallholders for outbreaks and failing to incorporate social determinants of epizootic spillover.[542] In much the same way, a red veganism of the North carries its own burden of green histories, among them antebellum slaveholders aiming for "ecological" plantations and closing the cotton cycle by forcing slaves to eat cottonseed oil.[543]

Perhaps more important, there is no reason to think meat production in and of itself need have negative ecological impacts. In fact, it can be part of ecological restoration and a keystone of poor pastoral livelihoods across swaths of the Third World.[544]

We know from the work of the geographer-veterinarian Diana Davis with the Aarib in southern Morocco that these pastoralists are expert managers of their animals and the range alike, and that banning grazing has in fact harmed the health of the rangeland, where animals and people alike flourish in non-equilibrium dynamics.[545] The best way to use these "highly variable arid environments is to amplify and facilitate pastoralists' mobility and to strengthen common property systems," building up on the lifeways and knowledge systems of the herders themselves. We likewise know from the work of sociologist Ricardo Jacobs in South Africa that urban slum dwellers live a dual life, as urban workers and as herders and livestock-keepers.[546] Such work is part-and-parcel of their daily social reproduction. On what grounds should researchers from the North demand the cessation of these activities and their replacement with lab meat?

Or, to take a third example, we could consider the buffalo of North America, which had long had a symbiotic relationship with the short grasslands of the Great Plains. In such ecosystems, buffalo were the "keystone herbivores within the Great Plains, sharing complex landscapes with other herbivores and predators for nearly 10,000 years."[547] Their constant feeding and grinding of manure, seed, and spare herbaceous matter underfoot historically ensured the ecological biodiversity of that environment and was the cause of the boggling richness of the black soil of the Plains.

As the Plains were "settled" by epochal primitive accumulation, the capitalist political ecology of the settlers displaced that of the Plains Indians, setting the stage for massive population destructions and colonial genocides. Later, the wheat planted on those fields was sold on world markets to undercut Third World agricultural systems, or fed to fatten up animals, all to the great profit of private corporations in the United States. While wheat and other commodities of the Green Revolution perhaps paradoxically have led to starvation, hunger, ecological wreckage, and the loss of peasant knowledge across the Third World, in what we might think would be an obvious symmetry, we hear no calls for banning cereal farming in total.[548]

Instead, researchers increasingly advocate restoring to the Great Plains the buffalo or other large herbivores that are capable of mimicking the grazing patterns of those extirpated animals.[549]

Elsewhere, the Gwich'in of Alaska subsist off caribou, and across the Sahel, millions of pastoralists survive off the production and sale of animals and meat, for their own consumption or tied into petty-commodity production.[550] Banning animal agriculture means banning animal agriculture in this world, and not in another world, which means, we should be clear, banning all the actual instances where people are engaged in animal agriculture. What should happen to the many millions of people whose modes of life are considered inappropriate?

A herd of examples stampedes to the horizon, but the point on that front is clear enough. Advocating intervening in the Global South and blithely demanding adopting capitalist technology in the name of a socialist Half-Earth, as does Vettese, who orders that it "must be from pasture that an eco-austere world will derive the land needed" for tree planting, is a form of "natural geo-engineering," developed according to specific values, specific devaluations, and pathological externalizations. These are not the no-brainers their advocates presume.[551] Compulsory veganism and lab meat, endorsed by prominent social democrat Green New Dealers, among them UPenn sociology prof Daniel Aldana Cohen, consents to the brute confiscation and erasure of peasant and pastoral particularisms in the name of "universal" ideals: rewilding Earth upon the bones of supposedly atavistic peoples poor and brown.[552]

Rampant "afforestation" sidesteps what the *Yale Environment 360* article Vettesse cites actually focuses on as a widely diverse array of natural carbon sequestration strategies that don't resort to the age-old colonial strategy of planting trees.[553]

In fact, in Ethiopia, the model country for tree planting's carbon absorption, non-native eucalyptus have caused tremendous damage to soil nutrients and water tables.[554] In other arenas, tree cover concusses biodiversity, as savanna wildebeests have the odd trait of failing to flourish in the forests planted by Harvard fiction writers.[555] Adding trees reduces fires, but fires have beneficial ecosystemic functions: they burn off the vegetation that casts shade over the ground-level of the landscape.[556] In that way, regular burns actually produce the grass upon which animals eat. Planting nice green trees hither and yon may end up killing all the antelope—quite an outcome for our colonial vegans.

In other artificial forest zones, streams and rivers have dried up and shrunk, precisely what is forecast to occur under global warming. Do we wish to adopt a political ecology that helps accelerate the present change in climate?

Where Tech Meats Capital

Lab meat is not a good idea even on its own biogeological grounds. It requires a massive amount of energy, and given that most agree that we need to reduce, not increase, Northern energy consumption, it makes little sense we would adopt a method of making food that depends exclusively on electricity. Initial studies show that making it low- or zero-carbon would require a misnamed clean energy, with at best less-dirty energy dependent upon mining nuclear and non-nuclear metals also producing pollutants and the impetus for land grabbing.[557]

Such meat also requires feedstock, a complex broth in which it grows. Presently, some are made using, of all things, fetal cow's blood.[558] So much for vegetarianism. Most also require massive bioreactors made of plastic, which would need to number in the tens of millions to supply a similar amount of meat as people currently consume. Plastic, of course, is another material- and energy-intensive material. More expensive than cow's blood is an unlabeled witch's brew of glucose, amino acids, vitamins, and minerals from industrial monocrop inputs. Again, not very energy-efficient and serving only as the next dumping grounds for many of the very inputs industrial meat now absorbs.

Finally, the technology reinforces relations of production to which red vegans declare they object, depending entirely upon venture capital angel investors, who see in the "innovation" a path to a new generation in massive profits.[559]

Here, again, we see a recurring feature, where "technology" is imagined as a neutral set of gewgaws, rather than summoned into being, as Marx described, in a specific form, by specific people, for a specific set of purposes.[560] Under capitalism, tech also arrives with a specific set of material needs, which are made possible only out of artificially depressed prices, including environmentally unequal exchange, just another way to loot anyone only peripherally connected to centers of capital, from the Global South to rural sacrifice zones in the U.S. and Europe, just outside these centers.[561] All in the name of progress.

We would suggest instead taking the lead from the international movement for food sovereignty, which is organized under the umbrella of La Via Campesina, as close to a Fifth International as exists in our world today. LVC took its cue from, among others, those who wrote the Wilderswil Declaration on Livestock Diversity:

> We will continue to further develop alternative research approaches and technologies that allow us to be autonomous and put control of genetic resources and livestock breeding in the hands of livestock keepers and other small-scale producers. And we will organize ourselves to conserve rare breeds. We are committed to fighting for our lands, territories and grazing pastures, our migratory routes, including trans-boundary routes. We will build alliances with other social movements with similar aims and continue to build international solidarity. We will fight for the rights of livestock keepers which include the right to land, water, veterinary and other services, culture, education and training, access to local markets, access to information and

decision making, that are all essential for truly sustainable live-stock production systems. We are committed to finding ways of sharing access to land and other resources with pastoralists, indigenous peoples, small farmers and other food producers according to equitable, but controlled, access.[562]

Livestock are more than thirsty meat bags and poultry more than an egg a day. For smallholders, animals are multifunctional, with a kaleidoscope of ecological and economic contributions.[563] They are stores of capital for communities that do not have easy access to banking systems. They are modes of transport. They work on fields and make labor that is backbreaking and tortuous possible. They eat forage from marginal and unplantable fields, and essentially work as protein farms with miraculous efficiency, gathering up photosynthetic energy converted to cellulose and turning it into meat. Amusingly, we do not need artificial (and unidimensional) meat incubators, since nature and the *longue durée* of human cultivation have provided the real deal for us.

Animals also poop, and manure directly enriches soil, restoring its nitrogen balance, providing a haven for soil organic matter, and generally producing beautifully rich and fertile soil perfect for farming. All without extracting almost the entirety of smallholder income for multinational chemical fertilizers (and other inputs) as occurs across so-called developed countries.[564]

For this reason, actual peasants—mysteriously absent from the TVD piece—have made very clear that they do not accept the termination of animal agriculture or compulsory veganism. Their demands are simple and clear, as in the resounding words of the Latin American Coordination of Rural Organizations (CLOC), a branch of La Via Campesina.[565] CLOC calls for "the promotion of peasant and indigenous family agriculture; a concept that encompasses all family-based agricultural activities, such as the way agriculture, livestock, forestry, fishing, aquaculture and grazing are organized, managed and operated by a family, and which depends on family labor."

Such a more-than-human community—extending beyond the family unit to broader landscapes—seems a much better option for the greater majority of the world than an Amtrak corridor–limited notion of ethics and appetites.

Nor are such proposals limited to the South. In the North, planned intensive rotations could sharply increase the Great Plains stocking capacity, at the same time increasing the quantity of animal per hectare and the quantity of carbon stocked away in the soil. Indeed, there are serious claims that meat in the long run could become carbon-negative, with knock-on effects that include increasing the capacity of the soil to retain water and its resilience in the face of the imminent or already present downpours of a warming world.

Smaller integrated farms are not merely a Southern peasant politic. Apparently unbeknownst to Muirian supremacist half-earthers at Harvard, they represent the core of a vibrant Northern food movement, wherein food sovereignty, indivisible from healthy soils, is undergoing a new renaissance, even in the face of agribusiness domination.

Here is a form of "natural geo-engineering" that we can get behind. Whether this would make meat more or less expensive, more or less available, we do not know, but when taking the perilous step of sketching out the cookshops of the future, the task at hand is to collaborate with the sustainable practices direct producers engage and to stick to non-negotiable demands such as unalienated production, ecological literacy, and egalitarianism. We need to avoid issuing blueprints for another world from the faculty dining club.

Left Aporophobia

There is no shortage of strange interventions. Rather than building up a programmatic post-COVID economics based on the living demands of movements in struggle, where, from the Philippines to Brazil, there are tightly disciplined mass rural movements calling for agrarian reform and agroecology, political economist Geoff Mann speaks *over* such movements—note the pattern—and advocates a new "experimental, adaptable and bold patchwork" that consists of "socializing" the food system.[566] How, then, ought it to be socialized?

The brief document to which he links is an odd patrician diatribe against food sovereignty, recycling wholly discredited arguments with which we authors have already dealt.[567] The only actual programmatic statement we could espy in the document to which Mann links argues *against* living wages for farmworkers and parity prices for farmers, the vanguard demands of the U.S. food sovereignty movement. Instead, True Socialists should promote automation of all "physically exhausting forms of toil" as a defense against the next pandemic.[568] The demand skips hand-in-hand with the techno-capitalist Breakthrough Institute regurgitating the long-debunked land sparing argument in favor of more intensive agriculture.[569] The demands are present in no food movement outside pro-agribusiness think-tank tax shelters arguing in favor of consolidating farms from underneath smallholders.[570]

Syracuse University geography professor Matt Huber, Mann's source, asks after what "automated technologies can be repurposed to create agroecological growing systems.... This means a debate based not on either industrial or smallholder agroecological production, but probably a combination of both."[571] One is left perplexed as to how Mann and Huber—marking themselves out of their depths—intend to impose agroecological growing systems with industrial production. Industrial farming refers to extensive capital inputs, whereas the question of the degree of mechanization of farming, with harvesters, for instance, is a topic upon which La Via Campesina is agnostic, leaving it up to farmers themselves.

Huber writes as if his objections to farmer autonomy are a matter of personal survival. The possibility farmers might choose to refuse him is a palpable panic—as if farmers aren't interested in feeding people!—recapitulating the two business parties' strategy in imposing divide-and-conquer upon rural and urban America.[572] Huber pays homage to the economies of scale, bourgeois central planning, and capitalism's sunk costs—tying relations of production to forces of production—that will sufficiently discipline producers and secure his larder. The tenured Kautskyist gone full Stalinist, leaving, as his *Jacobin* stablemates champion, chicken sandwiches for the plebs.[573]

The irony is that the next steps out of the agroeconomic traps that helped select for COVID-19, H5N2, and other outbreaks require making for a near-opposite heading. Not more of the same. Governmental intervention and regional planning *are* critical for helping agricultural communities emerging free from zones of agribusiness sacrifice, but decision-making in the spirit of the Zapatista principle of *mandar obecidiendo* (leadership from below) calls for those who best know how to grow food on *this,* the landscape they know so well, to help reinternalize a cycle of caring for the land generation-to-generation.[574]

The resulting virtuous cycles of regional food production—felt through land and labor alike—can be found all the way up through the geological scales and, as the International Panel on Sustainable Food Systems describe, the periurban food systems we all share:

> Wide-reaching shifts in social and economic relations also emerge as key components of agroecological transition. The Declaration of the International Forum on Agroecology states that "families, communities, collectives, organizations, and movements are the fertile soil in which agroecology flourishes. Solidarity between peoples, between rural and urban populations, is a critical ingredient."[575]

Summarizing a burgeoning literature, IPES-Food offers a program by which to rewire our food system *for all.* There are multiple examples of communities worldwide connecting ecological agriculture with urban markets, some operating at scales of millions of farmers and consumers.[576] Political agroecologist Jahi Chappell describes how Belo Horizonte, a city of 2.5 million people in Brazil, built a municipal food program that guaranteed a subsidized market of thousands in town for hinterland farmers, who could now afford agroecological and organic practices that protected local forests.[577]

To think that Huber calls himself a geographer, trafficking in the cheap divides of rural vs. urban and arguing food production has nothing to do with transportation. And if he insists on doing so— along the way citing Farshad Araghi as if the agrarian sociologist was in favor of depopulating the countryside rather than in appropriate repeasantization—it would be at best as a dishonest representative of a proud discipline.

Across even competing schools of agrarian studies, it's been long understood that for any movement on this front, we need to support farming communities' efforts to decide upon ecologically and socially sustainable levels of appropriate technology and mechanization. Given that we are currently in a race against time—in fact, we are in negative time—to produce the clean tech needed for keystone transition energy technologies, there seems no non-pathological reason to suggest using energy to mine, smelt, and work metals to build automated machines, which would increase U.S. society's overall energy use, making decarbonization harder, not easier.[578]

Better placed is figuring out ways that people might willingly accept and support the manual labor done by that brilliant machine for converting plant calories to mechanical energy, the human body. Alongside whatever automation farmers wish (as opposed to imposed by corporate end buyers or their paid scribes). Labor for labor, not for capital. Would that mean in the short term, double, triple, or ten times the minimum wage? We should all be for it!

But what do we find at the end of the line of citations, nested Russian doll-style, giving Mann plausible deniability, or, much more likely, signaling a simple lack of concern about the real-world consequences of the programmatic politics he espouses at his office? Nothing other than a propaganda piece for GMOs that prominently features plant geneticist Pamela Ronald, tied through more threads to chemical industry front groups such as the Cornell Alliance for Science (CAS)?[579]

Now, the "gotcha!" would be to wonder how *Viewpoint* magazine, where Mann's piece was published, and which in the past has published rigorous anti-Eurocentric work, came to launder the views of capitalist agribusiness and the chemical industry. But that would not take us very far, since we find the same with Vettesse, Taylor, and a Brooklyn loft party of kindred spirits. Each briefcase of sales brochures stashed in the proverbial cloakroom is redwashed as if such opportunism is all according to plan.

In Marx and Engels's names, Huber offers us that nothing is wrong with the present system save who runs it: "The goal of socialism is to take already existing socialized labor systems and socialize the control and benefits." Nah, bro, Marx vehemently disagreed.[580] Labor—its machines and ergonomics already capitalist impositions in relations of production—isn't the only source of wealth.[581] We have to take care of Earth too.

So placing monoculture plantations into worker control, as Huber demands, is neither the "ecological planning" he proposes on the one hand, nor, however necessary, a sufficient enough step in stopping pathogens from emerging out of the global circuits of production that the geographer weirdly also champions.

But such a fancy waves through proudly anti-rural cranks such as Doug Henwood. That *Left Business Observer,* sounding like Trump hawking hydroxychloroquine, recently posted CAS propaganda on his Facebook page about a "little-explored alternative" of delivering a COVID vaccine through genetically modified tomatoes. There's many a reason why it's "little-explored"—how, ironically enough, to assure standardized dosage?—but much as at John Ioannidis's Stanford, from Monsanto to the *Yankee Clipper* left, such ill-vetted, capital-led scientism runs express up the Northeast's coastline.

The problem is a more general one, beyond this particular terroir. Why are so many figures on the *bien pensant* Anglophone left adopting anti-ecological politics that advocate technologies that are as inseparable from their funders as the looms were from the mill owners in the age of the Luddites? Why are these positions serially platformed by allegedly critical podia, time and again, even as their logics are symmetrical to those underlying efforts to force meatpackers back to COVID-infested processing plants, where all that labor is "saved"? There's a through line from Trump to what counts in much of the Global North as the far left.

Clearly the interminable omission reflects an inability to center the voices of the actually existing ecological and anti-systemic movements in the core and periphery alike. Soul Fire Farms, the Savanna Institute, and the U.S. Food Sovereignty Alliance in the core are rendered invisible, as well as the more discomfiting and openly anti-imperialist La Via Campesina, which expresses solidarity with crucial fortresses for humanity's struggle for a better future such as Venezuela, Cuba, and the now-fallen Bolivia.[582]

Compare such calculated disappearances with the Minnesota Farmers Union's recent efforts to breach the rural-urban divide in the other direction:

> You've no doubt heard about the killing of George Floyd this week by a Minneapolis police officer. This horrific act and ensuing protests and property destruction have been hard to process, not just for those living and working in the Twin Cities Metro, but all Minnesotans and Americans.
>
> There's a lot to reckon with and soul-searching to do to ensure that, at an absolute minimum, nothing like this ever happens again. We have to do more than say that we condemn it, which we do. This comes on top of a deadly pandemic that has disproportionately harmed people of color, including in agriculture and food sectors.
>
> As always, we are here as a community, ready to listen to whatever is on your minds and hearts. Do not relegate this to simply an urban issue. We can't go back to the previous "normal" post-COVID—this makes it even clearer why. We call on our public officials to fight back against all injustices they can, and for everyone to reflect on why injustice persists.

Perhaps such soppy sentiments make us agroecologists "appreciate simplicity," not to say clarity, to borrow a condescending aphorism from a "radical" anatomy in favor of the Bolivian coup.[583] We leave that for others to judge, if in the glare of a bank of bright bulbs shining light therapy right to left upon the pandemic.

—MAY 30, 2020

To the Bat Cave

The tantric egg is the magma of all possibilities, the chaotic content looking for shape. The general intellect is the content, semio-capitalism is the gestalt, the generator of coded forms: paradigmatic capture. . . . Who will decide the actualization of one possibility or another?

—FRANCO BERARDI (2017)

I co-authored our finale for the first half of 2020—each week feeling a month long—with public health ecologist Deborah Wallace.

PEOPLE IN THE CITY OF Foshan in Guangdong province began falling ill with a disease that led to serious lung damage.[584] The illness was labeled, appropriately, Severe Acute Respiratory Syndrome, or SARS for short. In 2002–2003, SARS spread down the Kowloon Peninsula to Hong Kong and throughout China's southern provinces. It hopscotched internationally, with significant caseloads popping up in, among other places, Hanoi and Toronto, ultimately infecting over eight thousand people and killing 10 percent of those infected. The pathogen causing SARS was identified as part of a clade of coronavirus, the betacoronaviruses, different from the alphacoronaviruses, which cause some of the more severe common colds.

Epidemiologists in China went to work to find the source of this novel coronavirus.

Samples taken from wild animals on sale at the live market in Foshan yielded the virus in palm civets and a few other individual animals, including a single raccoon dog.[585] When these species were sampled in the wild, however, the SARS virus wasn't detected. Previous sampling of insectivorous bats had shown that they harbor a variety of viruses and research teams visited caves in the countryside.[586] Several bat species there had their guano sampled and, surely a fun time for all, their throats and anuses swabbed. The bats harbored not only a virus with very high consonance with the SARS genome, particularly the early human cases, but a slew of related coronaviruses that were dubbed "SARS-like" or SL.[587] Rhinolophid bats would also serve as a source for the Middle East Respiratory Syndrome (MERS)-CoV that emerged in Saudi Arabia a decade later.[588] Several milder human strains have also been identified: 229E, HKU1, NL63, and OC43.[589]

One of the species that most consistently harbored SARS and SL viruses was the Chinese horseshoe bat, or *Rhinolophus sinicus*.[590] This bat has a very wide range from Nepal through southern China and metapopulations large enough that it isn't considered endangered.[591] It lives mainly in caves and stays away from cities.[592] *R. sinicus* lives in groups of about a dozen to several hundred. Little has been published in English about its natural history.[593] The genus encompasses one species that is monogamous and one that is polyandrous, wherein only a few males in a colony mate with multiple females. The monogamous species is considered an oddity and most related species, many of whom also host SL, are assumed to be polyandrous.[594]

All bats in the genus *Rhinolophus* hibernate. Their reproductive cycles appear linked to both the hibernation and immunological status.[595] Reproduction's seasonality imprints upon "neuroimmunological reactivities" or how responsive the bat's neurons are to immunological intervention. So these reactivities aren't just a matter of some inherent properties of the different kinds of neurons. They are dependent in part upon environmental and social circumstances and combinations thereof.

In what way? Both males and females show a strong seasonality in gamete-producing tissue. The males' testes swell in late summer and produce sperm. The testes shrink in fall upon mating and hibernation. The females either store the sperm inseminated for the duration of the hibernation or store the zygotes that do not implant in the uterus until the other side of hibernation. Either possibility ensures that baby bats will be born during peak insect season, during which mothers and juveniles can feed amply on butterflies and beetles.[596] Along the way, the number of immunoreactive neurons in the brain declines in early June to late July, more so in males than females. That is, it's when male bats have their peak swelling and activity in the testes that their immunoreactive neurons decline. Trading off one head for another, as it were.

The seasonal biologies have been observed in different horseshoe bat species, but which one holds for the Chinese horseshoe bat has not been published in English. Indeed, the bat's very taxonomy remains to be worked out. Recent genetic analyses divide *Rhinolophus sinicus* into three subspecies: the widespread *R. s. sinicus,* the Yunnan Province–restricted *R. s. septentrionalis,* and the undescribed *R. s. ssp,* with the latter two now likely their own species.[597]

Our Inner Bat

Remarkably, as if a message from deep in our mammalian evolutionary history, the reproductive cycle of the horseshoe bat may be reflected in how SARS-1 and SARS-CoV-2, the second the virus behind the present COVID-19 pandemic, cause human infections, including their severity.

Once inside the human body, SARS latches onto the angiotensin-converting enzyme 2 (ACE2) receptor to gain entrance into the cells in which the virus replicates.[598] The target accounts for COVID-19 embodying as much a vascular disease and coagulopathy as a respiratory infection.[599] Researchers generally agree that levels of ACE2 in humans, distributed beyond lungs and across organ systems, influence the severity of illness and likelihood of death from COVID.[600]

It turns out that people born male have been hospitalized and have died of both SARS-1 and -2 at greater rates than those born female. And, to summarize an evolving and at times contradictory literature, males have higher concentrations of ACE2 in their blood and tissues.[601] Their testes are particularly rich in the receptor. The elderly also have higher concentrations of ACE2. Higher concentrations of the receptor also mark patients with certain chronic conditions, such as high blood pressure, coronary heart disease, and diabetes.[602] These starting conditions amplify disease impact together. Elderly biological men with chronic conditions die of COVID-19 at the greatest rates of all, a sacrifice out of *Midsommar* and favored by the lieutenant governor of Texas and other U.S. politicians with their eye on the state of the economy.[603]

A research group led by Carlos Wambier hypothesized direct connections among hormones, ACE2, and COVID outcome.[604] Upon an observation—and a follow-up hospital survey—that balding men appeared to be taking the worse of clinical courses, Wambier's team proposed a mechanism by which androgen load could impact COVID outcome. Transmembrane protease serine 2 both primes SARS-2's spike gene and may cleave ACE2, amplifying viral entry. Androgens aid in serine 2's transcription. As the genes for serine 2 and ACE2 are both located on the X chromosome, some clinical outcomes may also be determined in part by a genetic polymorphism. The team proposes androgen suppression as a possible treatment, with an in vitro strategy apparently successful in early work.[605]

The Wambier studies do not factor out age, which in and of itself can represent both disease susceptibility and balding. Other work has found that serine 2 expression in the lungs doesn't differ between sexes, with low levels of androgen in women enough to maintain serine 2.[606] The protease may also be regulated by estrogen. Clearly the impacts of hormones remain an unresolved matter.

A problem of a broader context, so-called sex hormones have such a command upon the societal (and the biomedical) imagination that, notwithstanding the plausibility of various proposed mechanisms, the variety of ways of conflating hormones and sex-specific outcomes is itself a subfield of study. "Scientific findings aren't served up on a platter by Mother Nature," observed sociomedical scientist Rebecca Jordan-Young and cultural anthropologist Katrina Karkazis of the science around testosterone:

> Instead, they are constructed out of specific research questions, the tools scientists use, and an enormous array of methodological choices, including what to measure and how, which groups or situations to compare, what statistical methods to use, and on and on. . . .
>
> We articulate the invisible brackets that ought to—but most often don't—appear at the end of statements about what [testosterone] does.... What sort of social relations does T enable or preclude in its particular manifestations? This question is not only relevant in the context of research; T is always embedded in social relations and other matters.[607]

Other groups are now including gender exposures shaping COVID outcomes that may extend beyond "biological" inputs. Medical researcher Catherine Gebhard and colleagues report a literature showing the comorbidities associated with COVID outcome—hypertension, cardiovascular disease, diabetes, even smoking—to be greater in men than women.[608] These gender norms appear also related to hand washing, mask wearing, and seeking health care.[609] On the other hand, greater ICU admissions for COVID in men, three- to fourfold more in Europe, may also be a function of medical access in men's favor. Caretaking professions meanwhile may place women in greater danger of infection and exposure inoculate. Or, in the other direction, as with MERS, men may be more exposed to source wild animal and livestock.[610] Could a heavy male bias in labor along various neoliberal agricultural frontiers serve as a selection window through which SL sex biases are amplified?[611]

How these various factors interact together and with more traditional physiological inputs, including differences in drug metabolism, remains to be investigated.[612] Elucidated, they could serve as points of intervention not only at the levels of the individual patient or hospital, but for policy and public health measures.

But given the difference in COVID incidences and clinical courses—and that sex-specific exposures regardless of their biosocial origins have been documented across diseases—following up on the bat-human analogue might tell us much about humanity's ill-explored place in the pathogen's evolutionary web. It should also help us think through the research routes scientists chose to take.

We learn, for instance, that human ACE2 and horseshoe bat ACE2 differ very little in their protein structures.[613] Follow-up experiments showed that *R. sinicus*'s ACE2 supported viral entry into bat-human chimeric cells by the SARS-CoV spike protein.[614] So humans and horseshoe bats—and other mammals, including mice—share common physiological "demands" or exposures.[615]

On the "supply" side, bats host an astonishing diversity of viruses.[616] Save a few pathogens and pests, white-nose syndrome among them, bats seem to show little effect from their microscopic passengers. Biologists have offered a variety of explanations that include the high basal temperatures, hibernation torpor, and ultra-strong immune resources that hold viruses to low levels for an organism that *must* be well enough to fly and survive.[617] Such immune virtuosity can select for virulence in pathogens upon spilling over into other species not so well-equipped.[618]

Virologists who sampled the Chinese horseshoe bat remarked especially on the variety of coronaviruses found in guano and swabs.[619] Few of these papers describe the bats sampled and their locale-specific conditions. Nor do the papers report when the samples were taken with respect to the hibernation and reproductive cycle.

Given that male bats have windows of vulnerability to viral damage in their reproductive cycle—during their bout of low immunoreactivity—more information is needed to address the possibility that pathogens help shape bat demographics and social evolution. Testing hypotheses that SARS and SL coronaviruses may play a role in shaping bat dynamics could help guide future research around the relationship between epizoology and social structure, with implications for the impact upon human disease outcomes. For instance, we might hypothesize that upon the end of seasonal competition for access to females, the virus serves as an epistatic way by which some of the excess males, especially those on the far end of their reproductive cycle, are weeded out to leave peak food for mothers and their offspring.

In polyandrous bats, low-status males play a role similar to drones in bee hives: something to get rid of when resources run low. Only a relatively few males are needed to impregnate a large number of females. The potential culling of males by viruses would occur in the swollen-testes phase of the reproductive cycle and ensure that only the most successful males survive to reproduce that season and beyond, until the male exits his reproductive prime.

Such sociobiology has been rigorously attacked when applied to humans. There's much more going on in humans than genetic profiles. Nature, on the other hand, is metal, to allude to an Instagram account of graphic videos of predation. Behavioral ecologies in nonhumans are often of a stern calculus. How is the distinction related to COVID outcome in humans? Clearly it's not that cis-women or childbearing trans men need to off their partners the way bats may. It's just that this is how SARS may have made its living coevolving with bats and once in humans merely rides a shared physiological mechanism to a different kind of epidemiological success.

Career FAO veterinarian Jan Slingenbergh and One Health consultant J. M. Leneman recently inventoried such shifts across an array of pathogens.[620] They conclude that the pathogenesis in a new livestock species or in humans reflects the life history of the virus in its original wild host, as we are proposing SARS-2 may do in humans.

Outer-body or epithelially transmitted infections, Slingenbergh and Leneman observe, tend toward recurrent spillovers of acute infections across a wide array of host taxa, largely using the same target tissue and modes of infection. Inner-body or more viscerally transmitted infections tend toward chronic infection of the deeper systems—such as immune or circulatory—of closely related host species. They are more likely to switch organ systems, including evolving vertical transmission from parent to offspring.

In reality, these disease types are arrayed across a continuum from the more epithelial infections such as the influenzas and coronaviruses spread by aerosols, direct contact, and particles on the wind, through the more visceral infections spread by fecal-oral, feed and water, venereal, vector-borne, and congenital, such as infectious bursal disease virus and bluetongue virus. In spite of attacking the lungs, which seem among our viscera, the SL coronaviruses are placed toward the epithelial pole because lung surface in effect offers an internalized skin. On ther other hand, SARS-2's attack across organs complicates the categorization.

These disease life histories also align with the opportunities humanity offers. Slingenbergh and Leneman propose growing global trade networks typically select for swiftly transmitted pathogens specific to the epithelial modes of transmission. Influenza's continuous attack, still underway today, began in the late nineteenth century with the establishment of commercial hog and poultry production.

What we add here is that such a cross-species imprint can extend to physiological traits that bats and humans share back to the Cretaceous-Paleogene boundary. Atop that, SARS may be redeploying the adaptations it developed to the behavioral ecologies of bat hosts toward solving its human targets in a new way.[621]

So what is good for the human goose is good for the viral gander. While we mine horseshoe crab blood for our own defense against a common bacterial enemy, as described earlier in the volume, SARS may be mining our symplesiomorphies, the ancestral character states we share with the horseshoe bat, to attack us in turn.[622] Deep evolution is a resource upon which animals across very different taxa may draw.

Where Was the Ecology in Ecohealth?

Why are the SARS suddenly attacking? Maybe we're just paying more attention. But there are reasons enough that such new diseases are emerging at a greater clip this century.[623]

China bears much, but not all, responsibility for the emergence of COVID-19. The BRICS approach aimed at using subimperial natural resources to those countries' own domestic advantage, raising millions of people out of poverty (while also leaving millions behind). The drawback included stripping dwindling forests and thereby increasing interfaces with previously marginalized epizootics, including SARS.[624]

The United States and Europe have played definitional roles themselves in funding the deforestation and development that this century led to the emergence of H1N1 (2009), Ebola Makona, H5N2, H5Nx, and Zika outside China. And, as we have discussed, helped produce many an influenza and SARS in China proper.[625] Indeed, in opposition to the mutual polarization of a new cold war, many of these outbreaks are joint productions.[626]

As sociologists Giovanni Arrighi and John Gulick put it, these competing centers of capital are foundationally integrated, sharing transnational supply lines, foreign direct investment, interlocking directorates, and a debt co-dependency during what appears a bout of common ecological collapse. Even the kind of dirty break from the United States to China (or any other formation) that world-systems theory predicts as next up in human history is no done deal.[627] There may be little environmental basis to support another iteration in global capital expansion.

Still, even with these complications, it's fascinating the way the United States squandered its advantage in imperial medicine in a matter of months, defunding the WHO and pursuing (a failure of) a nationalist program in COVID testing and vaccine development.[628] China meanwhile pivoted to fill the U.S. vacuum, sending doctors to NATO member Italy and millions of dollars in COVID aid to Africa.[629] China promised to make any vaccine it developed available to the world.[630]

The U.S. ideological apparatus hasn't given up, however. Indeed, it appears the new cold war has only begun. And not even just out of the Trumpist wing. We're talking Human Rights Watch, the *New York Times,* Lawfare, and CNN.[631] Google "China Africa COVID-19 donation" and see the battle rage for primacy on the internet high seas between these neoliberal outlets and China's media frigates, including *China Daily, Xinuanet,* and *Global Times.*[632] The headlines are illustrative. *Vox:* "How China is ruthlessly exploiting the coronavirus pandemic it helped cause."[633] *People's Daily:* "US politicians reveal their cold-bloodedness in pandemic response."[634]

The clash appears engulfing scientific domains, closing down open exchange of data and interpretation.

Reporting on the presence of a slew of coronaviruses in free-ranging bats in Myanmar, wildlife veterinarian Marc Valitutto and colleagues warned: "Ongoing land use change remains a prominent driver of zoonotic disease in Myanmar, bringing humans into ever closer contact with wildlife, and justifying continued surveillance and vigilance a broad scale."[635] Although the paper does not recommend reforms in land-use policies, it identifies one of the prominent upstream influences on transmission of viruses from bat to human. With rare exception, the scientific literature in China shies away from examining such social forces, including the roles played by agribusiness, the wild foods sector, and state-sponsored traditional medicine in breaching species disease barriers.

China's authorities have censored media and reviewed scientific papers submitted for publication, although likely to different extents in time and place. Government offices may not do this directly but expect heads of institutional departments to vet papers for potentially problematic political and economic implications. Recently, a policy came to light that indicates such vetting is happening during the pandemic.[636] The censorship this round focuses specifically on the origins of the outbreak in Wuhan to start and the global pandemic more generally. Sampling of bats began in earnest after the SARS-1 outbreak and was supported by teams that included wildlife biologists. Authorities were likely informed of the relationship between SARS and land use. The current vetting may simply formalize a policy that began with the first SARS outbreak, or even with earlier waves of avian influenza.

Boundaries between business and government in most countries have thinned considerably in recent decades, even beyond the already deep-seated class character of the state. In one egregious example, Margaret Hamburg, FDA Commissioner under Obama, refused to take the advice of agency experts on the growing opioid epidemic in the United States and continued to allow approval of new opioids for chronic pain. At that time, her husband was reaping millions in investments in the companies manufacturing opioids.[637] The lucrative damage of the opioid epidemic in the United States and the coronavirus outbreaks in China appear to mirror each other, including in the ways agribusiness land grabbing played decisive roles in both outcomes.[638]

Had they been made available, of what use would such advance warnings about land use and SARS have been for greater humanity?

If the millions of dollars spent on tracking the ecohealth of coronavirus in bats in alliance with scientists in China included more of the actual ecology underlying bat epizootics—and, as discussed earlier in the book, less of the dangerous gain-of-function experiments science studs are so gung-ho daring to dubious scientific gain—we might have been able to plot out the likely ramifications for a human spillover.[639]

Imagine commercials aimed at making dudes around the world practice proper public health. "This is Denny Hamlin," one might begin in the United States, "Even when I'm riding a hot engine at 210 miles per hour, I wear a mask to protect my pit crew. Back in my home town, I mask up for my friends and neighbors. We win some, we lose some, we wreck some. But let's leave our losses on the track. Put your mask on." Propagating gender roles in an effort to undercut a genderized health outcome offers pitfalls and ironies alike. But addressing a stubborn population in an emergency would allow us all to survive long enough to take up negotiating gender again—and, if successful, might help advance the campaign through a proverbial back door.

We've discussed mitigating land-use practices paid for by international capital that have increased the interface between bats and humans. There appears a related intervention that exploring the bat reproduction connection from the get-go might have helped us get started earlier. Conservation biologist Jiang Feng's group showed wide overlap in the trophic niches of R. *sinicus* and R. *affinis*—the intermediate horseshoe bat—differing somewhat in the prey species eaten, but not in prey size.[640] The overlap means there's plenty of prey to share. It begs the question why get rid of the male bats upon mating if food offers few limits?

But what if that's changed? Pesticide use expanded in China at the start of the century to four times that of a drenched United States, with 30 to 40 percent sprayed on cotton and new areas doused that map to some of the areas SL-hosting bats are located.[641] If pesticides have depressed beetle and butterfly populations, as suggested by a time series of insecticide use, taking off not long before SARS-1's emergence, would SARS strains that more sharply debilitated excess bat males be suddenly selected for? In the other direction, would reducing pesticide use mitigate virulence in animal pathogens capable of spilling over into humans and their livestock?

Which Possibility?

There is something of a wish fulfillment in the Denny Hamlin and pesticide interventions. As if our problem here is merely a matter of education or more information—the liberal panacea.

Unlike the celebrated board game, Pandemic in real life isn't merely a problem in combinatorial logistics that great minds can solve together. The challenge isn't just controlling a global problem with a contingent hand of cards. There are political economies actively manuveuring to block disease control if such efforts threaten vast fortunes.

In the global clash between polarities of accumulation we discussed, a bunch of big boulders are bouncing about past our heads, however bright our ideas. Not all us incorrigibles will professionally survive should we refuse one side of the clash or the other. Is another position even possible? Is another political regimen necessary before mapping out interventions into diseases brought about by a system entirely intent upon getting us back to "normal" and all its associated destruction?

There are a variety of third-campist stances. Some on the Anglo-American left embrace China as at the very least a Maoist counterpoise the country itself has repeatedly disavowed in word and deed.[642] Other leftists here have been accused of so denouncing China alongside the United States as if to signal domestic political officers that there is nothing to fear of them in their loyal opposition.[643] Even to the point that they platform CIA-funded assets. As there would be, denials abound.[644]

It begs whether there is a place for an internationalism outside a clash between turgid giants. Can such a formation pursue its own rules of engagement? Is that just wishful thinking born of growing desperation? Or does it call upon millions to gird their loins for a generational battle to wrest the ships of state from such alienated sociopathy? Do such questions only recapitulate the premises that got us into this mess?

That is, when we look into the bat cave does the cave look back into us? What else does it see we can't see about ourselves? What more did those epidemiologists—cultivated by their sponsors' interests—miss when looking down there in the dark these past twenty years? Does our future pitch us back more of the same?

At a high-end mall in Aventura, Florida, "COVID-19 Essentials," a pop-up store, recently opened to sell masks, hand sanitizers, UV light sterilizers, and shoe cover dispensers.[645] Not on the shelves, nor on offer by governments up and down jurisdiction in the United States: mass testing, contact tracing, community health visits, neighborhood food trucks, municipalized restaurants, monthly COVID checks, monthly mask and prophylaxis boxes mailed to each house, rent and mortgage abeyance, subsidized child care, and subsidized groceries from local farms.[646]

The racial disparities in COVID deaths in the States are worse than first thought.[647] The damage across the global divide is likely to take off, with the virus now full bore in Latin America and gearing up in Africa.[648] The just-in-time supply chains that produced capacity gaps in PPE, cleaning products, toilet paper, food, and other items, are financed in such a way that they are subjected to the kinds of debt-fueled market side bets that set off the housing crisis and Great Recession.[649] It's astonishing the lengths to which capital is willing to cannibalize itself!

The social costs are extending into the sociopsychological—particularly among isolated children—and the capitulation to a surveillance capitalism that in the States doesn't even offer the upside of apps that Taiwan and South Korea used to epidemiological advantage.[650]

Biomedical prospects remain hazy. Hundreds of labs around the world are on the hunt for an effective vaccine against SARS-2.[651] Between the bouts of misplaced optimism for and caustic pessimism against a vaccine touted ready in six months, there is the reality that no successful coronavirus vaccine has ever been produced.[652] It's truly a difficult problem. There is the matter that some vaccine efforts are skipping animal trials and jumping right to subjecting thousands of participants to experimentation, in some cases, even before a Phase 3 trial.[653] There is the danger that even an efficacious vaccine (or plasma therapy) might make matters worse in some recipients, from adverse reactions to driving antibody-dependent enhancement or Th2 immunopathology.[654] There is the likelihood of producing only partial protection, which others argue is better than no protection at all.

But even an efficacious vaccine faces an uphill battle. Trust is an epidemiological variable. Vaccine effectiveness is often a social negotiation.[655] Nearly half the United States is polling against taking a vaccine.[656] To get the percentage for herd immunity on board, making production gold standard should be strategized into the rollout, especially with hundreds of possible vaccines winging their way to market. Yes, time is of the essence. But "can we defuse the bomb?" may miss the point.

Meanwhile, the lovely human body may not be enough. Even many of those already infected appear to produce no neutralizing antibodies of their own.[657] And those that do so may be protected only for a few months.[658] The worry is that might mean those of us who got sick in March may already be vulnerable to another round of infection. On the other hand, T cell immunity against a wide array of SARS-2 targets appears effective, widely prevalent, and, upon exposure to the virus, long-term.[659] Yet, for many COVID cases that go south, T cells appear to be killed instead, as if SARS-2 were targeting them, although perhaps not directly in the way HIV does.

Not the best of prospects. Are there other futures instead? Can they bend us toward collectivities and practices beyond maintaining the present structure of power and accumulation?

As marvelous as biomedical interventions and the body can be, with vaccines and active antivirals now up in the air, should a public health of another metaphysics grab its moment? Nursing, for instance, is its own field, with a proud history, lively literature, and unique epistemology.[660] Successful hospitalization here isn't just about reductionist biology and the technology found in pill and push. It is also, often overwhelmingly, dependent upon the ergonomics of care that nurses administer. For lack of a better way of putting it, it is about the finely attuned *pratiques*—materially based rituals—around guiding a patient gutted by illness along the winding path to some semblance of wellness.

Can the nursing approach be scaled up across geography and domain? Can mutual aid be deployed in such a way that folk adaptation and parallel governance fill in for a state that has abandoned so many save the wealthy?[661] Can we hark back to a left tradition that found power not only in the place of work but in the neighborhoods in which workers lived?[662] Without a vaccine or a drug, many a country, officially operationalizing mutual aid efforts at scale, was able to drive its COVID incidences for all intents and purposes down to zero. Are the latter victories emblematic of Kojin Karantani's excavation of the rise of the state, which ended horizontal reciprocity as had been previously practiced, preserving mutual aid and equalization at the cost of the end of autonomy?[663] Is there room for both the state and localist cooperation?

Are we being revisited by still other eras? Masks—an intervention originating anywhere from the plague doctor of the Middle Ages to the Manchurian pneumonic plague epidemic of 1910–11—are definitively protective, especially in combination with other social practices, including physical distancing.[664] In the face of a flabby public ethos, the success of mask campaigns in the Global North requires legal enforcement.[665]

Are we being revisited by past human ecologies as well? Is there room for COVID detectors on a leash, their diagnoses more germane to an actual infection than temperature checks? Can dogs be deployed to detect COVID by smelling people's sweat?[666]

Are such nonhuman probiotic ecologies to be found farther out, in our barns, pastures, and buffer strips, where holistic interventions across regional landscapes can maneuver pathogens into parts of their parameter space where only the less virulent strains can win?[667] Are such efforts dependent on tying together degrowth, decolonization, developmental convergence, and debt cancellation, returning disalienation of land and labor front and center after five hundred years of getting quashed by capitalism?[668]

Our colleague, agrarian sociologist Max Ajl, sums it up:

> It is by now well-known that industrialized capitalist agriculture is an incubator for deadly viruses. Industrial farming is discredited, and agro-ecological movements should press their advantage. Across the world, ecological peasant movements, farmers' movements, permaculture initiatives braided with de-commodification of farmland, are building up the physical production systems for a resilient and non-capitalist world food system.
>
> Peasant movements are not distinct but are part-and-parcel of the ecological movement and the movement of the laboring classes. They are therefore a sterling example of the environmentalism of the poor. Such movements [in conjunction with a pro-people State] are the basis for global just transitions, as they can build up food production systems which feed the poor, restore the health of the planet, and are excellent prophylaxis against viruses.[669]

New futures out of an old and tired present. But suddenly other projectiles whiz by our ear from wrong angles out of the bat cave's depths.

A genetics team led by biodiversity researcher Shu-Miaw Chaw, for instance, just proposed the bat-pangolin recombinant that served as the progenitor for SARS-2, described earlier in the book, in actuality emerged *forty years ago*.[670] And with their source strains' shared amino acids in the receptor-binding domain maintained by purifying selection ever since, SARS-2, the team hypothesizes, has been circulating undetected in humans in China *years* before Wuhan. Early SARS-2 may not have had the obvious affinity for human-to-human infection of the strain that emerged in mid-December and increasingly so in the months to follow, for instance, as represented by the amino acid replacement D614G in the spike protein.[671]

More surprises. The EcoHealth Alliance, putting the ecology back into its genetics studies, sampled 334 Sunda pangolins confiscated in Malaysia from 2009–2019, finding no coronaviruses in the upstream of the pangolin trade.[672] Let's set aside the study's statistical power for the moment—22,000 pangolins of a much larger trade were logged by a single illegal operation in Sabah between 2007 and 2009 alone.[673] The EcoHealth Alliance team argues that the infected batch of pangolins in Guangdong in 2019, whose coronavirus matches with SARS-2's receptor binding domain, may have been an incidental infection. There are also significant issues with the variety of papers reporting pangolin sequences, including designating different sequences which appear instead from the same source.[674] At this time, the original 2019 pangolin isolate still appears legit, even as Chaw's results indicate the recombination event needn't depend on an actual pangolin host.

Another projectile, the next generation of the lab hypothesis, streaks by. It moves from declaring a bat coronavirus sampled in 2013 escaped one of the two Wuhan labs to proposing it was samples taken from six miners that came down with COVID-like symptoms in 2012—and became adapted to humans—that escaped.[675] An interesting line of investigation worth pursuing, and yet, despite what seem compounding coinkydinks, we remain unconvinced.[676]

This version of the lab hypothesis suffers from:

- conflating genetic similarity plots and phylogenetic inference;
- confusing outgroups for direct ancestors;
- dismissing documented recombination frequencies in wild-type SLs;
- claiming the structure of the spike protein in the 2013 bat sample converged on SARS-2's spike, but skipping over the still closer structure of the 2019 pangolin sample;
- claiming the six miners *may* have selected for the polybasic furin cleavage sites in SARS-2, but omitting the actual furin sites found in wild-type SLs;
- rejecting the possibility of human adaptation across millions of animals in the field (including multitudes of incidental humans), but presuming six miners enough to condense decades of SARS evolution in a matter of weeks;
- claiming the comparatively little variation in the genetics of pandemic SARS-2 a marker of lab origins when that represents no such proof and is more likely a marker of the speed of the spread of the virus (even as such variation is also starting to accumulate); and
- dancing around an out-and-out accusation that the cover-up of a lab accident must now extend beyond the Wuhan labs and to scientists working as far afield as Beijing, Shandong, Guangdong, and Yunnan.

Proponents of the lab hypothesis claim parisomony's mantle. It seems the explanation offers no such thing.

Well, if anything, all the topsy-turvy going on around characterizing COVID-19 is most decidedly disconcerting. And only six months into the pandemic, we shouldn't be surprised. While the Chaw interpretation does speak to the spatiotemporal scope of SL origins we proposed, what began in mystery lurches onward in mystery.[677] Our book here offers a COVID map that at best appears rough, but, we'd argue, one far better than that which landed us in this horror.

As Ajl and his roundtable co-participants put it starkly, capitalism, in whichever country it is practiced, is most definitely *not* metal.[678] There's nothing necessary in its cruelty. Multinational agribusiness, mining, logging, and real estate chop at the tree of life for greed first and foremost, the planet and its people bedamned. Pleased at the ruins it surveys, the system, tended to by dead epidemiologists, served up COVID-19 on its proverbial arm. Upon this perch, the virus now leisurely eats through humanity alive.

—JULY 24, 2020

Notes

Preface

1. Wallace R, Y-S Huang, P Gould, and D Wallace (1997). "The hierarchical diffusion of AIDS and violent crime among U.S. metropolitan regions: Inner-city decay, stochastic resonance and reversal of the mortality transition." *Social Science & Medicine* 44(7): 935–947; Jhung MA, D Swedlow, SJ Olsen, D Jergnigan, M Biggerstaff, et al. (2011). "Epidemiology of 2009 pandemic Influenza A (H1N1) in the United States." *Clinical Infectious Diseases* 52(S1): S13-S26; Wallace RG (2020). "Setting the jet net." Patreon, 5 July. https://www.patreon.com/posts/setting-jet-net-38980079.
2. Desjardins MR, A Hohl, and EM Delmelle (2020). "Rapid surveillance of COVID-19 in the United States using a prospective space-time scan statistic: Detecting and evaluating emerging clusters." *Applied Geography* 118: 102202; Amin R, T Hall, J Church, D Schlierf, and M Kulldorff (2020). "Geographical surveillance of COVID-19: Diagnosed cases and death in the United States." medRxiv, 25 May. https://www.medRxiv.org/content/10 .1101/2020.05.22.20110155v1.
3. Maxmen A (2020). "The race to unravel the biggest coronavirus outbreak in the United States." *Nature*, 6 March. https://www.nature.com/articles/ d41586-020-00676-3.
4. Akuno K and A Nangwaya (2017). *Jackson Rising: The Struggle for Economic Democracy and Black Self-Determination in Jackson, Mississippi*. Daraja Press; Minka A (2020). "The self-confessed bankruptcy of Mayor Chokwe Antar Lumumba." *Black Agenda Report*, 1 July. https://blackagendareport. com/self-confessed-bankruptcy-mayor-chokwe-antar-lumumba.
5. Gilpin L (2018). "'We don't have any voice': Rural Mississippians feel shut out, overcharged by electric co-ops." *Mississippi Today*, 14 December. https://mississippitoday.org/2018/12/14/we-dont-have-any-voice-rural

-mississippians-feel-shut-out-overcharged-by-electric-co-ops/; Fairchild DG (2020). "Powering democracy through clean energy." In D Orr, A Gumbel, B Kitwana, and W Becker (eds), *Democracy Unchained: How to Rebuild Government for the People*. The New Press, New York. Dawson A (2020). *People's Power: Reclaiming the Energy Commons*. OR Books, New York.

Notes on a Novel Coronavirus

6.   Ramadan N and H Shaib (2019). "Middle East respiratory syndrome coronavirus (MERS-CoV): a review." *Germs* 9(1): 35–42.

7.   Anonymous (2020). "Coronavirus: Death toll climbs, and so does the number of infections." *New York Times*, 28 January. https://www.nytimes.com/2020/01/28/world/asia/china-coronavirus.html.

8.   Buckley C, R Zhong, D Grady, and RC Rabin (2020). "As coronavirus fears intensify, effectiveness of quarantines is questioned." *New York Times*, 26 January. https://www.nytimes.com/2020/01/26/world/asia/coronavirus-wuhan-china-hubei.html.

9.   Read JM, JRE Bridgen, DAT Cummings, A Ho, and CP Jewell (2020). "Novel coronavirus 2019-nCoV: Early estimation of epidemiological parameters and epidemic predictions." medRxiv, 28 January. https://www.medRxiv.org/content/10.1101/2020.01.23.20018549v2.article-info.

10. The Nextstrain Team (2020). "Genomic epidemiology of novel coronavirus." https://nextstrain.org/ncov/global.

11. Center for Systems Science and Engineering at Johns Hopkins University (2020). "COVID-19 dashboard." https://gisanddata.maps.arcgis.com/apps/opsdashboard/index.html#/bda7594740fd40299423467b48e9ecf6.

12. Griffiths J (2020). "Number of Wuhan coronavirus cases inside mainland China overtakes SARS, as virus spreads worldwide." *CNN*, 28 January. https://www.cnn.com/2020/01/28/asia/wuhan-coronavirus-update-jan-29-intl-hnk/index.html.

13. Wallace R and RG Wallace (2016). "The social amplification of pandemics and other disasters." In Wallace R and RG Wallace (eds), *Neoliberal Ebola: Modeling Disease Emergence from Finance to Forest and Farm*. Springer International Publishing, Cham, pp 81–93.

14. Szabo L (2020). "Something far deadlier than the Wuhan coronavirus lurks near you, right here in America." *USA Today*, 24 January. https://www.usatoday.com/story/news/health/2020/01/24/coronavirus-versus-flu-influenza-deadlier-than-wuhan-china-disease/4564133002/.

15. Defoe D (1995). *A Journal of the Plague Year*. Project Gutenberg EBook. https://www.gutenberg.org/files/376/376-h/376-h.htm; Barry J (2005). *The Great Influenza*. Penguin Books, New York.

16. Dawood F, et al. (2012). "Estimated global mortality associated with the first 12 months of 2009 pandemic influenza A H1N1 virus circulation: a modelling study." *The Lancet Infectious Diseases* 12(9): 687–695.

17. Johnson C (2020). "Scientists are unraveling the Chinese coronavirus with

unprecedented speed and openness." *Washington Post*, 24 January. https://www.washingtonpost.com/science/2020/01/24/scientists-are-unraveling-chinese-coronavirus-with-unprecedented-speed-openness/.

18. Kapczynski DR, MJ Sylte, ML Killian, MK Torchetti, and K Chrzastek (2017). "Protection of commercial turkeys following inactivated or recombinant H5 vaccine application against the 2015 U.S. H5N2 clade 2.3.4.4 highly pathogenic avian influenza virus." *Veterinary Immunology and Immunopathology* 191: 74–79.

19. Quan-Cai C, et al. (2006). "Refined estimate of the incubation period of severe acute respiratory syndrome and related influencing factors." *American Journal of Epidemiology* 163(3): 211-216.

20. Cohen E (2020). "China says coronavirus can spread before symptoms show—calling into question US containment strategy." *CNN*, 26 January. https://www.cnn.com/2020/01/26/health/coronavirus-spread-symptoms-chinese-officials/index.html.

21. Ashton, JR (2003). "Type I and type II errors exist in public health practice too." *Journal of Epidemiology and Community Health* 57: 918.

22. Wallace RG, H HoDac, R Lathrop, and W Fitch (2007). "A statistical phylogeography of influenza A H5N1." *Proceedings of the National Academy of Sciences* 104 (11): 4473–4478; Wallace RG (2016). *Big Farms Make Big Flu.* Monthly Review Press, New York.

23. Wallace RG (2017). "Prometheus rebound." *Farming Pathogens* blog, 31 October. https://farmingpathogens.wordpress.com/2017/10/31/prometheus-rebound/.

24. Wallace RG (2013). "The bug has left the barn." *Farming Pathogens* blog, 15 January. https://farmingpathogens.wordpress.com/2013/01/15/the-bug-has-left-the-barn/.

25. Wallace RG (2013). "Broiler explosion." *Farming Pathogens* blog, 14 April. https://farmingpathogens.wordpress.com/2013/04/14/broiler-explosion/.

26. Wallace RG, et al (2016). "Did neoliberalizing West African forests produce a new niche for Ebola?" *International Journal of Health Services.* 46(1): 149–165.

27. Branswell H (2019). "The data are clear: Ebola vaccine shows 'very impressive' performance in outbreak." *STAT*, 12 April. https://www.statnews.com/2019/04/12/the-data-are-clear-ebola-vaccine-shows-very-impressive-performance-in-outbreak; Kupferschmidt K (2019). "Finally, some good news about Ebola: Two new treatments dramatically lower the death rate in a trial." *Science*, 12 August. https://www.sciencemag.org/news/2019/08/finally-some-good-news-about-ebola-two-new-treatments-dramatically-lower-death-rate.

28. Vogel C, et al. (2019). "Clichés can kill in Congo." *Foreign Policy*, 30 April. https://foreignpolicy.com/2019/04/30/cliches-can-kill-in-congo-grand-nord-north-kivu-tropes-conflict-ebola-response/.

29. Davis M (2018). *Old Gods, New Enigmas.* Verso Press, New York.

30. Myers, SL (2020). "China's omnivorous markets are in the eye of a lethal

outbreak once again." *New York Times*, 25 January. https://www.nytimes. com/2020/01/25/world/asia/china-markets-coronavirus-sars.html.

31. Ji W, W Wang, X Zhao, J Zai, and X Li (2020). "Cross-species transmission of the newly identified coronavirus 2019-nCoV." *Journal of Medical Virology* 92(4): 433–440.

32. Xinhua (2020). "China detects large quantity of novel coronavirus at Wuhan seafood market." *Xinhuanet*, 27 January. http://www.xinhuanet. com/english/2020-01/27/c_138735677.htm.

33. Huang C, et al (2020). "Clinical features of patients infected with 2019 novel coronavirus in Wuhan, China." *The Lancet* 395(10223): 497–506.

34. Cohen J (2020). "Wuhan seafood market may not be source of novel virus spreading globally." *Science*, 26 January. https://www.sciencemag.org/ news/2020/01/wuhan-seafood-market-may-not-be-source-novel-virus-spreading-globally.

35. Bradsher K and A Tang (2019). "China responds slowly, and a pig disease becomes a lethal epidemic." *The New York Times*, 17 December. https:// www.nytimes.com/2019/12/17/business/china-pigs-african-swine-fever. html.

36. Wallace R, Bergmann L, Hogerwerf L, Kock R, and RG Wallace (2016). "Ebola in the hog sector: Modeling pandemic emergence in commodity livestock." In Wallace R and RG Wallace (eds), *Neoliberal Ebola: Modeling Disease Emergence from Finance to Forest and Farm*. Springer International Publishing, Cham, pp 13–53.

37. Wallace RG (2007). "The great bird flu name game." H5N1 blog post, 27 December. https://mronline.org/wpcontent/uploads/2020/01/rg_wallace_ the_great_bird_flu_name_game.pdf.

38. Anonymous (2020). "Coronavirus death toll climbs in China, and a lockdown widens." *The New York Times*, 23 January. https://www.nytimes. com/2020/01/23/world/asia/china-coronavirus.html.

39. Wallace RG, L Bergmann, L Hogerwerf, and M Gilbert (2010). "Are influenzas in Southern China byproducts of the region's globalizing historical present?" In J Gunn, T Giles-Vernick, and S Craddock (eds), *Influenza and Public Health*. Routledge Press, London; Zhong T, et al. (2018). "The impact of proximity to wet markets and supermarkets on household dietary diversity in Nanjing city, China." *Sustainability* 10(5): 1465.

40. Broglia A and C Kapel (2011). "Changing dietary habits in a changing world: Emerging drivers for the transmission of foodborne parasitic zoonoses." *Veterinary Parasitology* 182(1): 2-13; Liu Q, L Cao, and X Zhu (2014). "Major emerging and re-emerging zoonoses in China: a matter of global health and socioeconomic development for 1.3 billion." *International Journal of Infectious Diseases* 25: 65–72; Schneider M (2017). "Wasting the rural: Meat, manure, and the politics of agro-industrialization in contemporary China." *Geoforum* 78: 89–97.

41. Wallace RG, L Bergmann, L Hogerwerf, and M Gilbert (2010). "Are

influenzas in Southern China byproducts of the region's globalizing historical present?"

42. Dzoma BM, S Sejoe, and BVE Segwagwe (2008). "Commercial crocodile farming in Botswana." *Tropical Animal Health and Production* 40: 377–381; Brooks EGE, SI Roberton, and DJ Bell (2010). "The conservation impact of commercial wildlife farming of porcupines in Vietnam." *Biological Conservation.* 143(11): 2808–2014; Mather C and A Marshall (2011). "Living with disease? Biosecurity and avian influenza in ostriches." *Agriculture and Human Values* 28: 153–165; Kamins AO, et al. (2011). "Uncovering the fruit bat bushmeat commodity chain and the true extent of fruit bat hunting in Ghana, West Africa." *Biological Conservation* 144(12): 3000–3008; Yulia M and D Suhandy (2017). "Indonesian palm civet coffee discrimination using UV-visible spectroscopy and several chemometrics methods." *Journal of Physics Conference Series.* https://iopscience.iop.org/article/10.1088/1742-6596/835/1/012010/meta#artAbst.

43. Roach J (2011). "New shark species found in food market." *National Geographic,* September 11. https://www.nationalgeographic.com/news/2011/9/110901-shark-new-species-eaten-science-ocean-squalus-formosus-dogfish/.

44. Foster, J B and B Clark (2009). "The paradox of wealth: Capitalism and ecological destruction." *Monthly Review* 61(6). https://monthlyreview.org/2009/11/01/the-paradox-of-wealth-capitalism-and-ecological-destruction/.

45. Fearnley L (2013). "The birds of Poyang Lake: Sentinels at the interface of wild and domestic." *Limn* 3. https://limn.it/issues/sentinel-devices/.

46. Wallace, RG (2019). "Review of Paul Richards' *Ebola: How a People's Science Ended an Epidemic*." *Antipode Online,* 13 May. https://antipodeonline.org/2019/05/13/ebola-how-a-peoples-science-helped-end-an-epidemic/.

47. Zhang, QF (2012). "The political economy of contract farming in China's agrarian transition." *Journal of Agrarian Change* 12(4): 460–483; Wallace R (2014). "Collateralized farmers." *Farming Pathogens* blog, 8 May. https://farmingpathogens.wordpress.com/2014/05/08/collateralized-farmers/.

48. Wallace RG (2018). *Duck and Cover: Epidemiological and Economic Implications of Ill-Founded Assertions that Pasture Poultry Are an Inherent Disease Risk.* Australian Food Sovereignty Alliance. https://mronline.org/wp-content/uploads/2020/01/Wallace-Duck-and-Cover-Report-September-2018.pdf; Okamoto K, A Liebman, and RG Wallace (2019). "At what geographic scales does agricultural alienation amplify foodborne disease outbreaks? A statistical test for 25 U.S. states, 1970–2000." *medRxiv,* 8 January. https://www.medRxiv.org/content/10.1101/2019.12.13.19014910v2.

49. Okamoto K, A Liebman, and RG Wallace (2019). "At what geographic scales does agricultural alienation amplify foodborne disease outbreaks? A statistical test for 25 U.S. states, 1970–2000."

50. Wallace RG (2010). "We can think ourselves into a plague." *Farming Pathogens*

blog, 25 October. https://farmingpathogens.wordpress.com/2010/10/25/we-can-think-ourselves-into-a-plague/.

51. LePageM(2020). "CRISPR-editedchickensmaderesistanttoacommonvirus." *New Scientist*, 20 January. https://www.newscientist.com/article/2230617-crispr-edited-chickens-made-resistant-to-a-common-virus/amp/.

52. Wallace RG (2010). "The Scientific American." *Farming Pathogens* blog, 18 January. https://farmingpathogens.wordpress.com/2011/01/18/the-scientific - american/.

53. Wallace RG, et al. (2015). "The dawn of Structural One Health: a new science tracking disease emergence along circuits of capital." *Social Science and Medicine* 129: 68–77.

54. Reid S (2020). "How the coronavirus started in China—and why that's actually a saving grace." *ABC News*, 23 January. https://www.abc.net.au/news/2020-01-24/coronavirus-came-from-animals-stopping-spread-not-simple/11893420.

55. Press A (2019). "On the origins of the professional-managerial class: an interview with Barbara Ehrenreich." *Dissent*, 22 October. https://www.dissentmagazine.org/online_articles/on-the-origins-of-the-professional-managerial-class-an-interview-with-barbara-ehrenreich.

56. Huber M (2019). "Ecological politics for the working class." *Jacobin*, 12 October. https://jacobinmag.com/2019/10/ecological-politics-working-class-climate-change.

57. Wallace RG (2016). *Big Farms Make Big Flu*.

58. Lazare S (2016). "Ultra-rich 'philanthrocapitalist' class undermining global democracy: report." *Common Dreams*, 15 January. https://www.commondreams.org/news/2016/01/15/ultra-rich-philanthrocapitalist-class-undermining-global-democracy-report.

59. Greenfeld KT (2009). *China Syndrome*. HarperCollins, New York.

60. Wallace RG et al. (2014). "Did Ebola emerge in West Africa by a policy-driven phase change in agroecology?" *Environment and Planning A* 46: 2533-2542; Wallace RG (2015). "Made in Minnesota." *Farming Pathogens* blog, 10 June. https://farmingpathogens.wordpress.com/2015/06/10/made-in-minnesota; Wallace RG (2017). "Industrial production of poultry gives rise to deadly strains of bird flu H5Nx." Institute for Agriculture and Trade Policy blog, 24 January. https://www.iatp.org/blog/201703/industrial-production-poultry-gives-rise-deadly-strains-bird-flu-h5nx; Wallace R, et al. (2018). *Clear-Cutting Disease Control: Capital-Led Deforestation, Public Health Austerity, and Vector-Borne Infection*. Springer International Publishing, Cham.

61. Wallace RG (2009). "The hog industry strikes back." *Farming Pathogens* blog entry, 1 June. https://farmingpathogens.wordpress.com/2009/06/01/the-hog-industry-strikes-back; Wallace, RG (2015). "Made in Minnesota."

62. Foster JB (2013). "Marx and the rift in the universal metabolism of

nature." *Monthly Review* 65(7). https://monthlyreview.org/2013/12/01/
marx-rift-universal-metabolism-nature/.

63. Wallace RG (2016). *Big Farms Make Big Flu.*

Interview: "Agribusiness Would Risk Millions of Deaths"

64. Pabst Y (2020) "Coronavirus: 'Agribusiness would risk millions of deaths.'"
    Marx21, 11 March. https://www.marx21.de/coronavirus-agribusiness-
    would-risk-millions -of-deaths/.
65. De Crescenzo L (2020) "Pandemic strike." *Uneven Earth*, 16 March. http://
    unevenearth.org/2020/03/pandemic-strike/.

COVID-19 and Circuits of Capital

66. Roser M, H Ritchie, and E Ortiz-Ospina (2020). "Coronavirus disease
    (COVID-19)—statistics and research." *Our World in Data*, 22 March. https://
    ourworldindata.org/coronavirus#growth-country-by-country-view.
67. Rosenthal BM, Goldstein J, and Rothfeld M (2020). "Coronavirus in N.Y.:
    'Deluge' of cases begins hitting hospitals." *New York Times*, 20 March.
    https://www.nytimes.com/2020/03/20/nyregion/ny-coronavirus-hospitals.
    html.
68. Rappleye H, AW Lehren, L Stricklet, and S Fitzpatrick (2020). "'The system
    is doomed': Doctors, nurses, sound off in NBC News coronavirus survey."
    *NBC News, 20 March. https://www.nbcnews.com/news/us-news/system-
    doomed-doctors-nurses-sound-nbc-news-coronavirus-survey-n1164841.*
69. Relman E (2020). "The federal government outbid states on critical
    coronavirus supplies after Trump told governors to get their own medi-
    cal equipment." *Business Insider*, 20 March. https://www.businessinsider.
    com/coronavirus-trump-outbid-states-on-medical-supplies-2020-3;
    Oliver D (2020). "Trump announces U.S.-Mexico border closure to stem
    spread of coronavirus." *USA Today*, 19 March. https://www.usatoday.com/
    story/travel/news/2020/03/19/u-s-mexico-officials-look-ban-non-essen-
    tial-travel-across-border/2874497001/.
70. Ferguson N, et al. on behalf of the Imperial College COVID-19 Response
    Team (2020). "Impact of Non-Pharmaceutical Interventions (NPIs) to
    reduce COVID-19 mortality and healthcare demand." 16 March. https://
    spiral.imperial.ac.uk:8443/handle/10044/1/77482.
71. Taleb NN (2007). *The Black Swan.* Random House, New York; Shen C,
    NN Taleb, and Y Bar-Yam (2020). "Review of Ferguson et al. 'Impact of
    Non-Pharmaceutical Interventions.'" *New England Complex Systems
    Institute*, 17 March. https://necsi.edu/review-of-ferguson-et-al-impact
    -of-non-pharmaceutical-interventions.
72. NewTmrw (2020). "Coronavirus is too radical…" Twitter, 21 March.
    https://twitter.com/NewTmrw/status/1241532936760909825.
73. Wallace R (2020). "Pandemic firefighting vs. pandemic fire prevention."
    Unpublished manuscript, 20 March. Available upon request.

74. Allen J (2020). "Trump's not worried about coronavirus: but his scientists are." *NBC News*, 26 February. https://www.nbcnews.com/politics/white-house/trump-s-not-worried-about-coronavirus-his-scientists-are-n1143911; Riechmann R (2020). "Trump disbanded NSC pandemic unit that experts had praised." *AP News*, 14 March. https://apnews.com/ce014d94b64e98b7203b873e56f80e9a.

75. Sanger DE, E Lipton, E Sullivan, and M Crowley (2020). "Before virus outbreak, a cascade of warnings went unheeded." *New York Times*, 19 March. https://www.nytimes.com/2020/03/19/us/politics/trump-coronavirus-outbreak.html.

76. Taylor M (2020). "Exclusive: U.S. axed CDC expert job in China months before virus outbreak." *Reuters*, 22 March. https://www.reuters.com/article/us-health-coronavirus-china-cdc-exclusiv/exclusive-us-axed-cdc-expert-job-in-china-months-before-virus-outbreak-idUSKBN21910S.

77. Waitzkin H (ed) (2018). *Health Care Under the Knife: Moving Beyond Capitalism for Our Health*. Monthly Review Press, New York.

78. Lewontin R and R Levins (2000). "Let the numbers speak." *International Journal of Health Services* 30(4): 873–77.

79. Matthews O (2020). "Britain drops its go-it-alone approach to coronavirus." *Foreign Policy*, 17 March. https://foreignpolicy.com/2020/03/17/britain-uk-coronavirus-response-johnson-drops-go-it-alone; Wallace R (2020). "Pandemic strike." *Uneven Earth*, 16. March http://unevenearth.org/2020/03/pandemic-strike; Frey I (2020). " 'Herd immunity' is epidemiological neoliberalism." *Quarantimes*, 19 March. https://thequarantimes.wordpress.com/2020/03/19/herd-immunity-is-epidemiological-neoliberalism/.

80. Payne A (2020). "Spain has nationalized all of its private hospitals as the country goes into coronavirus lockdown." *Business Insider*, 16 March. https://www.businessinsider.com/coronavirus-spain-nationalises-private-hospitals-emergency-covid-19-lockdown-2020-3.

81. Lange J (2020). "Senegal is reportedly turning coronavirus tests around 'within 4 hours' while Americans might wait a week." *Yahoo News*, 12 March. https://news.yahoo.com/senegal-reportedly-turning-coronavirus-tests-165224221.html.

82. Sterling S and JM Morgan (2019). *New Rules for the 21st Century: Corporate Power, Public Power, and the Future of Prescription Drug Policy in the United States*. Roosevelt Institute, New York.

83. Koebler J (2020). "Hospitals need to repair ventilators: Manufacturers are making that impossible." *Vice*, 18 March. https://www.vice.com/en_us/article/wxekgx/hospitals-need-to-repair-ventilators-manufacturers-are-making-that-impossible.

84. Wang M, et al. (2020). "Remdesivir and chloroquine effectively inhibit the recently emerged novel coronavirus (2019-nCoV) in vitro." *Cell Research* 30: 269–271.

85. Anonymous. (2020). "Autonomous groups are mobilizing mutual aid

initiatives to combat the coronavirus" *It's Going Down*, 20 March. https://itsgoingdown.org/autonomous-groups-are-mobilizing-mutual-aid-initiatives-to-combat-the-coronavirus/.

86. Andersen K, A Rambaut, WI Lipkin, EC Holmes, and RF Garry (2020). "The proximal origin of SARS-CoV-2." *Nature Medicine, 17 March. https://www.nature.com/articles/s41591-020-0820-9.*

87. Wallace RG. "Notes on a novel coronavirus." This volume.

88. Gilbert M, et al. (2020). "Preparedness and vulnerability of African countries against importations of COVID-19: a modelling study." *Lancet* 395(10227): 871–877.

89. Sun J (2015). "The regulation of 'novel food' in China: the tendency of deregulation." *European Food and Feed Law Review* 10(6): 442–448.

90. Brooks EGE, SI Roberton, and DJ Bell (2010). "The conservation impact of commercial wildlife farming of porcupines in Vietnam." *Biological Conservation* 143(11): 2808–2014.

91. Schneider M (2017). "Wasting the rural: Meat, manure, and the politics of agro-industrialization in contemporary China." *Geoforum* 78: 89–97.

92. Wallace RG, L Bergmann, L Hogerwerf, and M Gilbert (2010). "Are influenzas in Southern China byproducts of the region's globalizing historical present?" In J Gunn, T Giles-Vernick, and S Craddock (eds), *Influenza and Public Health: Learning from Past Pandemics.* Routledge, London; Broglia A and C Kapel (2011). "Changing dietary habits in a changing world: Emerging drivers for the transmission of foodborne parasitic zoonoses." *Veterinary Parasitology* 182(1): 2–13.

93. Jones KE, NG Patel, MA Levy, A Storeygard, D Balk, et al. (2008). "Global trends in emerging infectious diseases." *Nature* 451(7181): 990–993; Molyneux D, et al. (2011). "Zoonoses and marginalised infectious diseases of poverty: Where do we stand?" *Parasites & Vectors* 4(106). https://parasitesandvectors.biomedcentral.com/articles/10.1186/1756-3305-4-106.

94. Morse SS, et al. (2012). "Prediction and prevention of the next pandemic zoonosis." *Lancet* 380(9857): 1956–1965; Wallace RG (2016). *Big Farms Make Big Flu.* Monthly Review Press, New York.

95. Wallace RG, et al. (2015). "The dawn of Structural One Health: a new science tracking disease emergence along circuits of capital." *Social Science and Medicine* 129: 68–77.

96. Cummins S, S Curtis, AV Diez-Roux, S Macintyre (2007). "Understanding and representing 'place' in health research: a relational approach." *Social Science & Medicine* 65(9): 1825–1838; Bergmann L and M Holmberg (2016). "Land in motion." *Annals of the American Association of Geographer* 106(4): 932–956; Bergmann L (2017). "Towards economic geographies beyond the nature-society divide." *Geoforum* 85: 324–335.

97. Jorgenson AK (2006). "Unequal ecological exchange and environmental degradation: A theoretical proposition and cross-national study of deforestation, 1990–2000." *Rural Sociology* 71(4): 685–712; Mansfield B, Munroe

DK, and McSweeney K (2010). "Does economic growth cause environmental recovery? Geographical explanations of forest regrowth." *Geography Compass* 4(5): 416–427; Hecht SB (2014). "Forests lost and found in tropical Latin America: the woodland 'Green Revolution.'" *Journal of Peasant Studies* 41(5): 877–909; Oliveira GLT (2016). "The geopolitics of Brazilian soybeans." *Journal of Peasant Studies* 43(2): 348–72.

98. Turzi M (2011). "The soybean republic." *Yale Journal of International Affairs* 6(2); Haesbaert R (2011). *El Mito de la Desterritorialización: Del 'Fin de Los Territorios' a la Multiterritorialidad*. Siglo Veintiuno, Mexico City; Craviotti C (2016). "Which territorial embeddedness? Territorial relationships of recently internationalized firms of the soybean chain." *Journal of Peasant Studies* 43(2): 331–347.

99. Jepson W, C Brannstrom, and A Filippi (2010). "Access regimes and regional land change in the Brazilian Cerrado, 1972–2002." *Annals of the Association of American Geographers* 100(1): 87–111; Meyfroidt P, et al. (2014). "Multiple pathways of commodity crop expansion in tropical forest landscapes." *Environmental Research Letters* 9(7); Oliveira GLT (2016). "The geopolitics of Brazilian soybeans"; Godar J (2016). "Balancing detail and scale in assessing transparency to improve the governance of agricultural commodity supply chains." *Environmental Research Letters* 11(3).

100. Wallace R, et al. (2018). *Clear-Cutting Disease Control: Capital-Led Deforestation, Public Health Austerity, and Vector-Borne Infection*. Springer International Publishing, Cham.

101. Davis M (2016). *Planet of Slums*. Verso, New York; Moench M and Gyawali D (2008). *Desakota: Reinterpreting the Urban-Rural Continuum*. Institute for Social and Environmental Transition, Kathmandu; Hecht SB (2014). "Forests lost and found in tropical Latin America: the woodland 'Green Revolution.'"

102. Lugo AE (2009). "The emerging era of novel tropical forests." *Biotropica* 41(5): 589–591.

103. Wallace R and RG Wallace (eds) (2016). *Neoliberal Ebola: Modeling Disease Emergence from Finance to Forest and Farm*. Springer International Publishing, Cham; Wallace R et al (2018). *Clear-Cutting Disease Control: Capital-Led Deforestation, Public Health Austerity, and Vector-Borne Infection*; Kallis G and E Swyngedouw (2018). "Do bees produce value? A conversation between an ecological economist and a Marxist geographer." *Capitalism Nature Socialism* 29(3): 36–50.

104. Wallace RG, et al. (2016). "Did neoliberalizing West African forests produce a new niche for Ebola?" *International Journal of Health Services*. 46(1): 149–165.

105. Wallace R and RG Wallace (eds) (2016). *Neoliberal Ebola: Modeling Disease Emergence from Finance to Forest and Farm*.

106. Bicca-Marques JC and D Santos de Freitas (2010). "The role of monkeys, mosquitoes, and humans in the occurrence of a yellow fever outbreak in

a fragmented landscape in South Brazil: Protecting howler monkeys is a matter of public health." *Tropical Conservation Science* 3(1): 78–89; Bicca-Marques JC, et al. (2017). "Yellow fever threatens Atlantic forest primates." *Science Advances* e-letter, 25 May; Oklander LI (2017). "Genetic structure in the southernmost populations of black-and-gold howler monkeys (*Alouatta caraya*) and its conservation implications." *PLoS ONE* 12(10); Fernandes NCCA et al (2017). "Outbreak of yellow fever among nonhuman primates, Espirito Santo, Brazil, 2017." *Emerging Infectious Diseases* 23(12): 2038–2041; Mir D (2017). "Phylodynamics of yellow fever virus in the Americas: New insights into the origin of the 2017 Brazilian outbreak." *Scientific Reports* 7(1).

107. Davis M (2005). *The Monster at Our Door: The Global Threat of Avian Flu.* New Press, New York; Graham JP, et al. (2008). "The animal-human interface and infectious disease in industrial food animal production: Rethinking biosecurity and biocontainment." *Public Health Reports* 123(3): 282–299; Jones BA, et al. (2013). "Zoonosis emergence linked to agricultural intensification and environmental change." *PNAS* 110(21): 8399–8404; Liverani M, et al. (2013). "Understanding and managing zoonotic risk in the new livestock industries." *Environmental Health Perspectives* 121(8); Engering A, L Hogerwerf, and J Slingenbergh (2013). "Pathogen-host-environment interplay and disease emergence." *Emerging Microbes and Infections* 2(1); Slingenbergh J (ed) (2013). *World Livestock 2013: Changing Disease Landscapes.* Food and Agriculture Organization of the United Nations, Rome.

108. Tauxe RV (1997). "Emerging foodborne diseases: an evolving public health challenge." *Emerging Infectious Diseases* 3(4): 425–434; Wallace R and RG Wallace, eds (2016) *Neoliberal Ebola: Modeling Disease Emergence from Finance to Forest and Farm*; Marder EP (2018). "Preliminary incidence and trends of infections with pathogens transmitted commonly through food—foodborne diseases active surveillance network, 10 U.S. sites, 2006–2017." *Morbidity and Mortality Weekly Report* 67(11): 324–328.

109. Wallace RG (2009). "Breeding influenza: the political virology of offshore farming." *Antipode* 41(5): 916–951; Wallace RG, et al. "The origins of industrial agricultural pathogens." This volume.

110. Vandermeer JH (2011). *The Ecology of Agroecosystems.* Jones and Bartlett, Sudbury, MA; Thrall PH, et al. (2011). "Evolution in agriculture: the application of evolutionary approaches to the management of biotic interactions in agro-ecosystems." *Evolutionary Applications* 4(2): 200–215; Denison RF (2012). *Darwinian Agriculture: How Understanding Evolution Can Improve Agriculture.* Princeton University Press, Princeton; Gilbert M, X Xiao, and TP Robinson (2017). "Intensifying poultry production systems and the emergence of avian influenza in China: A 'One Health/Ecohealth' epitome." *Archives of Public Health* 75:48.

111. Houshmar M, et al. (2012). "Effects of prebiotic, protein level, and stocking

density on performance, immunity, and stress indicators of broilers." *Poultry Science* 91(2): 393–401; Gomes AVS et al. (2014). "Overcrowding stress decreases macrophage activity and increases salmonella enteritidis invasion in broiler chickens." *Avian Pathology* 43 (1): 82–90; Yarahmadi P, HK Miandare, S Fayaz, and CMA Caipang (2016). "Increased stocking density causes changes in expression of selected stress- and immune-related genes, humoral innate immune parameters and stress responses of rainbow trout (*Oncorhynchus mykiss*)." *Fish & Shellfish Immunology* 48: 43–53; Li W, et al. (2019). "Effect of stocking density and alpha-lipoic acid on the growth performance, physiological and oxidative stress and immune response of broilers." *Asian-Australasian Journal of Animal Studies* 32(12).

112. Pitzer VE, et al. (2016). "High turnover drives prolonged persistence of influenza in managed pig herds." *Journal of the Royal Society Interface* 13(119): 20160138; Gast RK, et al. (2017). "Frequency and duration of fecal shedding of *Salmonella* Enteritidis by experimentally infected laying hens housed in enriched colony cages at different stocking densities." *Frontiers in Veterinary Science* 61(3): 366–371; Diaz A (2017). "Multiple genome constellations of similar and distinct Influenza A viruses co-circulate in pigs during epidemic events." *Scientific Reports* 7: 11886.

113. Atkins KE, et al. (2011). "Modelling Marek's disease virus (MDV) infection: Parameter estimates for mortality rate and infectiousness." *BMC Veterinary Research* 7: 70; Allen J and Lavau S (2015). "'Just-in-time' disease: Biosecurity, poultry and power." *Journal of Cultural Economy* 8(3): 342–360; Pitzer VE, et al. (2016). "High turnover drives prolonged persistence of influenza in managed pig herds"; Rogalski MA (2017). "Human drivers of ecological and evolutionary dynamics in emerging and disappearing infectious disease systems." *Philosophical Transactions of the Royal Society B* 372(1712): 20160043.

114. Wallace RG (2009). "Breeding influenza: the political virology of offshore farming"; Atkins KE, et al. (2013). "Vaccination and reduced cohort duration can drive virulence evolution: Marek's disease virus and industrialized Agriculture." *Evolution* 67(3): 851–860; Mennerat A, MS Ugelvik, CH Jensen, and A Skorping (2017). "Invest more and die faster: The life history of a parasite on intensive farms." *Evolutionary Applications* 10(9): 890–896.

115. Nelson MI, et al. (2011). "Spatial dynamics of human-origin H1 Influenza A virus in North American swine," *PLoS Pathogens* 7(6): e1002077; Fuller TL (2013). "Predicting hotspots for influenza virus reassortment." *Emerging Infectious Diseases* 19(4): 581–588; Wallace R and RG Wallace (2014). "Blowback: New formal perspectives on agriculturally-driven pathogen evolution and spread." *Epidemiology and Infection* 143(10): 2068–2080; Mena I, et al. (2016). "Origins of the 2009 H1N1 influenza pandemic in swine in Mexico." *eLife* 5: e16777; Nelson MI, et al. (2019). "Human-origin influenza A(H3N2) reassortant viruses in swine, Southeast Mexico." *Emerging Infectious Diseases* (25)4: 691–700.

116. Wallace RG (2016). "The dirty dozen" in *Big Farms Make Big Flu*. Monthly Review Press, New York, pp 192–201.

117. Centers for Disease Control and Prevention (2015). *Safer Food Saves Lives*. 3 November. https://www.cdc.gov/vitalsigns/foodsafety-2015/index. html; Sun, LH (2015). "Big and deadly: Major foodborne outbreaks spike sharply." *Washington Post*, 3 November. https://www.washingtonpost. com/news/to-your-health/wp/2015/11/03/major-foodborne-outbreaks-in-u-s-have-tripled-in-last-20-years; Stobbe, M (2015). "CDC: More food poisoning outbreaks cross state lines." *KSL, 3 November*. https://www.ksl.com/article/37217542/cdc-more-food-poisoning-outbreaks-cross-state-lines.

118. Goldenberg S (2018). "Alicia Glen, who oversaw de Blasio's affordable housing plan and embattled NYCHA, to depart city hall." *Politico*, 19 December. https://www.politico.com/states/new-york/city-hall/story/2018/12/19/alicia-glen-who-oversaw-de-blasios-affordable-housing-plan-and-embattled-nycha-to-depart-city-hall-760480.

119. Dymski GA (2009). "Racial exclusion and the political economy of the subprime crisis." *Historical Materialism* 17: 149–179; Barnett HC (2011). "The securitization of mortgage fraud." *Sociology of Crime, Law and Deviance* 16: 65–84.

120. Ivry B, B Keoun, and P Kuntz (2011). "Secret Fed loans gave banks $13 billion undisclosed to Congress." *Bloomberg*, 21 November. https://www.bloomberg.com/news/articles/2011-11-28/secret-fed-loans-undisclosed-to-congress-gave-banks-13-billion-in-income.

121. Merced MJ and D Barboza (2013). "Needing pork, China is to buy a U.S. supplier." *New York Times*, 29 May. https://www.nytimes.com/2020/03/18/us/politics/china-virus.html.

122. SCMP Reporter *(2008)*. "Goldman Sachs pays US$300m for poultry farms." *South China Morning Post,* 4 August. https://www.scmp.com/article/647749/goldman-sachs-pays-us300m-poultry-farms.

123. 5m Editor (2008). "Goldman Sachs invests in Chinese pig farming." *Pig Site*, 5 August. https://thepigsite.com/news/2008/08/goldman-sachs-invests-in-chinese -pig-farming-1.

124. Rogers K, L Jakes, and A Swanson (2020). "Trump defends using 'Chinese virus' label, ignoring growing criticism." *New York Times*, 18 March. https://www.nytimes.com/2020/03/18/us/politics/china-virus.html.

125. Marx K (1894; 1993). *Capital: A Critique of Political Economy*, vol. 3. Penguin: New York, p 362.

126. Lipton E, N Fandos, S LaFraniere, and JE Barnes (2020). "Stock sales by Senator Richard Burr ignite political uproar." *New York Times*, 20 March. https://www.nytimes.com/2020/03/20/us/politics/coronavirus-richard-burr-insider-trading.html.

127. Mossavar-Rahmani S, et al. (2020). *ISG Insight: From Room to Grow to Room to Fall*. Goldman Sachs' Investment Strategy Group, 16 March. https://www.goldmansachs.com/insights/pages/from-room-to-grow-to-room-to-fall.html.

128. Anonymous *(2020)*. "Corona crisis: Resistance in a time of pandemic." *marx21*, 21 March. https://www.marx21.de/corona-crisis-resistance-in-a-time-of-pandemic; International Assembly of the Peoples and Tricontinental Institute for Social Research (2020). "In light of the global pandemic, focus attention on the people. *Tricontinental*, 21 March. https://www.thetricontinental.org/declaration-covid19/.

129. Wallace RG, et al. (2015). "The dawn of Structural One Health: a new science tracking disease emergence along circuits of capital."

130. Wallace RG, et al. (2016). "Did neoliberalizing West African forests produce a new niche for Ebola?" *International Journal of Health Services*; Wallace R, et al. (2018). *Clear-Cutting Disease Control: Capital-Led Deforestation, Public Health Austerity, and Vector-Borne Infection.*

131. Mandel E (1970). "Progressive disalienation through the building of socialist society, or the inevitable alienation in industrial society?" In E Mandel, *The Marxist Theory of Alienation.* Pathfinder, New York; Virno P (2004). *A Grammar of the Multitude.* Semiotext(e), Los Angeles; Weston D (2014). *The Political Economy of Global Warming: The Terminal Crisis.* London: Routledge; Wark M (2017). *General Intellects: Twenty-One Thinkers for the Twenty-First Century.* Verso, New York; Foster JB (2018). "Marx, value, and nature." *Monthly Review* 70(3): 122–136; Federici S (2018). *Re-enchanting the World: Feminism and the Politics of the Commons.* PM, Oakland.

132. Lee B and Red Rover (1993). *Night-Vision: Illuminating War and Class on the Neo-Colonial Terrain.* Vagabond, New York; Federici, S (2004). *Caliban and the Witch: Women, the Body and Primitive Accumulation.* Autonomedia, New York; Tsing, A (2009). "Supply chains and the human condition." *Rethinking Marxism* 21(2): 148–176; Coulthard, GS (2014). *Red Skin, White Masks: Rejecting the Colonial Politics of Recognition.* University of Minnesota Press, Minneapolis; Vergara-Camus, L (2014). *Land and Freedom: The MST, the Zapatistas and Peasant Alternatives to Neoliberalism.* Zed, London; Wang J (2018). *Carceral Capitalism.* Semiotext(e), Los Angeles.

133. Haraway D (1991). "A cyborg manifesto: Science, technology, and socialist-feminism in the late Twentieth Century." In D Haraway, *Simians, Cyborgs and Women: The Reinvention of Nature.* Routledge, New York; Taylor KY (2017). *How We Get Free: Black Feminism and the Combahee River Collective.* Haymarket, Chicago.

134. Fracchia J (2017). "Organisms and objectifications: a historical-materialist inquiry into the 'Human and the animal.'" *Monthly Review* 68(10): 1–17; Giraldo OF (2019). *Political Ecology of Agriculture: Agroecology and Post-Development.* Basel: Springer, Cham.

135. Berardi F (2009). *The Soul at Work: From Alienation to Autonomy.* Semiotext(e), Los Angeles; Lazzarato M (2014). *Signs and Machines: Capitalism and the Production of Subjectivity.* Semiotext(e), Los Angeles; Wark M (2017). *General Intellects: Twenty-One Thinkers for the Twenty-First Century.*

136. Wallace R, A Liebman, L Bergmann, and Wallace RG (2020). "Agribusiness

vs. public health: Disease control in resource-asymmetric conflict." https://
hal.archives-ouvertes.fr/hal-02513883.

137. Wallace RG, K Okamoto, and A Liebman (2020). "Gated ecologies." In DB
Monk and M Sorkin (eds), *Between Catastrophe and Revolution: Essays in
Honor of Mike Davis*. Terreform/OR Books, New York.

138. Wallace R, et al. (2018). *Clear-Cutting Disease Control: Capital-Led
Deforestation, Public Health Austerity, and Vector-Borne Infection.*

139. Wallace RG, et al. "The origins of industrial agricultural pathogens." This
volume.

Interview: "Internationalism must sweep away globalization"

140. Anonymous (2020) "Internationalism must sweep away globalization."
Jabardakhal, 8 April. http://jabardakhal.in/english/internationalism-must-
sweep-away -globalization-rob-wallace/.

141. Oliver D (2020). "Trump announces U.S.-Mexico border closure to stem
spread of coronavirus." *USA Today,* 19 March. https://www.usatoday.com/
story/travel/news/2020/03/19/u-s-mexico-officials-look-ban-non-essen-
tial-travel-across-border/2874497001/.

142. Davis M (2020). "Who gets forgotten in a pandemic." *The Nation,* 13 March.
https://www.thenation.com/article/politics/mike-davis-covid-19-essay/.

143. Spinney L (2017). "Who names diseases?" *Aeon,* 23 May. https://aeon.co/
essays/disease-naming-must-change-to-avoid-scapegoating-and-politics;
Wallace RG. "Midvinter-19." This volume.

144. Feldman A (2020). "States bidding against each other pushing up prices
of ventilators needed to fight coronavirus, NY Governor Cuomo says."
Forbes, 28 March. https://www.forbes.com/sites/amyfeldman/2020/03/28/
states-bidding-against-each-other-pushing-up-prices-of-ventilators-
needed-to-fight-coronavirus-ny-governor-cuomo-says/#76a38b00293e;
Artenstein AW (2020). "In pursuit of PPE." *New England Journal of Medicine,*
30 April. https://www.nejm.org/doi/full/10.1056/NEJMc2010025.

145. Brown M (2020). "Illinois adjusts on the fly to meet medical supply needs
in a coronavirus 'Wild West.'" *Chicago Sun-Times,* 3 April. https://chi-
cago.suntimes.com/coronavirus/2020/4/3/21207488/coronavirus-illinois
-medical-supplies-wild-west.

146. Holmes K, C Hassan, and D Williams (2020). "New England Patriots
team plane with 1.2 million N95 masks arrives from China to help ease
shortages." CNN, 3 April. https://www.cnn.com/2020/04/02/us/coronavi-
rus-patriots-plane-masks-spt-trnd/index.html.

147. Arrighi G (2009). *Adam Smith in Beijing: Lineages of the 21st Century.*
Verso, New York.

148. Poggioli S (2020). "For help on coronavirus, Italy turns to China, Russia
and Cuba." NPR, 25 March. https://www.npr.org/sections/coronavirus-
live-updates/2020/03/25/821345465/for-help-on-coronavirus-italy-turns
-to-china-russia-and-cuba.

149. Lange J (2020). "Senegal is reportedly turning coronavirus tests around 'within 4 hours' while Americans might wait a week." *Yahoo News*, 12 March. https://news.yahoo.com/senegal-reportedly-turning-coronavirus-tests-165224221.html.

150. Sui C (2020). "In Taiwan, the coronavirus pandemic is playing out very differently. What does life without a lockdown look like?" *NBC News*, 23 April. https://www.nbcnews.com/news/world/taiwanese-authorities-stay-vigilant-virus-crisis-eases-n1188781.

151. Anonymous (2020). "NY governor says China is donating 1,000 ventilators to the state." CNN, 4 April. https://www.cnn.com/world/live-news/coronavirus-pandemic-04-04-20/h_1f631979b35b4bcfa05223e75c101e9b.

152. Fitz D (2020). "How Che Guevara taught Cuba to confront COVID-19." *Monthly Review*, 1 June. https://monthlyreview.org/2020/06/01/how-che-guevara-taught -cuba-to-confront-covid-19/.

The Kill Floor

153. Johns Hopkins Coronavirus Resource Center (2020). "COVID-19 dashboard by the Center for Systems Science and Engineering (CSSE)." 9 June 2020. https://coronavirus.jhu.edu/map.html.

154. Reynolds E and H Pettersson (2020). "Confirmed coronavirus cases are rising faster than ever." CNN, 5 June. https://www.cnn.com/2020/06/05/world/coronavirus-cases-rising-faster-intl/index.html.

155. Correa A (2020). "5 mil indígenas estão com covid-19 na Região Pan Amazônica." *GQ*, 3 June. https://gq.globo.com/Corpo/Saude/noticia/2020/06/5-mil-indigenas-estao-com-covid-19-na-regiao-pan-amazonica.html; Otis J (2020). "The coronavirus is spreading through indigenous communities in the Amazon." NPR, 12 June. https://www.npr.org/2020/06/12/873091962/coronavirus-hits-indigenous-groups-in-colombian-amazon-on-brazilian-border.

156. Thebault R and A Hauslohner (2020). "A deadly 'checkerboard': Covid-19's new surge across rural America." *Washington Post*, 24 May. https://www.washingtonpost.com/nation/2020/05/24/coronavirus-rural-america-outbreaks/.

157. Ibid.

158. Nguyen M (2020). "File:COVID-19 outbreak USA stay-at-home order county map.svg." Wikimedia Commons, 22 March. https://commons.wikimedia.org/wiki/File: COVID-19_outbreak_USA_stay-at-home_order_county_map.svg.

159. Owen P (2020). "Giuliani calls COVID-19 contact tracing 'ridiculous': 'We should trace everybody for cancer' (Video)." *The Wrap*, 23 April. https://www.thewrap.com/giuliani-calls-covid-19-contact-tracing-ridiculous -we-should-trace-everybody-for-cancer-video/; Feldman N (2020). "America has no plan for the worst-case scenario on Covid-19." Bloomberg, 6 May. https://finance.yahoo.com/news/america-no-plan-worst-case-153036385.

html; Flaxman S, S Mishra, A Gandy, HJT Unwin,TA Mellan, et al. (2020). "Estimating the effects of non-pharmaceutical interventions on COVID-19 in Europe." *Nature*, https://doi.org/10.1038/s41586-020-2405-7; Hsiang S, D Allen, S Annan-Phan, K Bell, I Bolliger, et al. (2020). "The effect of large-scale anti-contagion policies on the COVID-19 pandemic." *Nature*, https://www.nature.com/articles/s41586-020-2404-8.

160. Scheuber A and SL van Elsland (2020). "Potential US COVID-19 resurgence modelled as lockdowns ease." Imperial College London, 21 May. https://www.imperial.ac.uk/news/197656/potential-us-covid-19-resurgence-modelled-lockdowns/; Mervosh S, JC Lee, L Gamio, and N Popovich (2020). "See how all 50 states are reopening." *New York Times*, June 10. https://www.nytimes.com/interactive/2020/us/states-reopen-map-coronavirus.html.

161. Czachor A (2020). "Meatpacking giant closes South Dakota plant 'indefinitely' after almost 300 employees test positive for COVID-19." *Newsweek*, 12 April. https://www.newsweek.com/meatpacking-giant-closes-south-dakota-plant -indefinitely-after-almost-300-employees-test-positive-1497498.

162. Chadde S (2020) "Tracking COVID-19's impact on meatpacking workers and industry." Midwest Center for Investigative Reporting, 16 April. https://investigatemidwest.org/2020/04/16/tracking-covid-19s-impact -on-meatpacking-workers-and-industry/.

163. Graddy S, S Rundquist, and B Walker (2020). "Investigation: Counties with meatpacking plants report twice the national average rate of COVID-19 infections." *EWG News and Analysis*, 14 May. https://www.ewg.org/news-and-analysis/2020/05/ewg-map-counties-meatpacking-plants-report-twice-national-average-rate.

164. Mulvany L, J Skerritt, P Mosendz, and J Attwood (2020). "Scared and sick, U.S. meat workers crowd into reopened plants." Bloomberg, 21 May. https://www.bloomberg.com/news/articles/2020-05-21/scared-and-sick-u-s-meat-workers -crowd-into-reopened-plants.

165. Moody K (2020). "How 'just-in-time' capitalism spread COVID-19." *Spectre*, 8 April. https://spectrejournal.com/how-just-in-time-capitalism-spread-covid-19/; Clapp J (2020) "Spoiled milk, rotten vegetables and a very broken food system." *New York Times*, 8 May.

166. Lin X, PJ Ruess, L Marston, and M Konar (2019). "Food flows between counties in the United States." *Environmental Research Letters*, 26 July. https://iopscience.iop.org/article/10.1088/1748-9326/ab29ae.

167. Ibid. Konar M (2019). "We mapped how food gets from farms to your home." *The Conversation*, 25 October. https://theconversation.com/we-mapped-how-food-gets-from-farms-to-your-home-125475.

168. Gorsich EE, RS Miller, HM Mask, C Hallman, K Portacci, and CT Webb (2019). "Spatio-temporal patterns and characteristics of swine shipments in the U.S. based on Interstate Certificates of Veterinary Inspection." *Scientific Reports*, 9(1). https://www.nature.com/articles/s41598-019-40556-z.

169. Ibid.
170. Swanson A, D Yaffe-Bellany and M Corkery (2020). "Pork chops vs. people: Battling coronavirus in an Iowa meat plant." *New York Times*, 1o May. https://www.nytimes.com/2020/05/10/business/economy/coronavirus-tyson-plant-iowa.html; Driver A (2020). "Arkansas poultry workers amid the coronavirus: 'We're not essential, we're expendable'." *Arkansas Times*, 11 May. https://arktimes.com/arkansas-blog/2020/05/11/arkansas-poultry-workers-amid-the-coronavirus-were-not-essential-were-expendable.
171. Sinclair U (1906; 2015). *The Jungle*. Dover Publications; Kotz N (1967). "Meat industry abuse revealed in USDA probe." *Minneapolis Tribune*, 16 July; Genoways T (2014). *The Chain: Farm, Factory, and the Fate of our Food*. Harper, New York.
172. Coleman J (2020). "Meatpacking worker told not to wear face mask on job died of coronavirus: report." *The Hill*, 7 May. https://thehill.com/policy/finance/496595-meatpacking-worker-told-not-to-wear-face-mask-on-job-died-of-coronavirus; Stueese A (2020). "He worked for better conditions at his chicken plant. Then the coronavirus took his life." *Clarion Ledger*, 12 May. https://www.clarionledger.com/story/opinion/voices/2020/05/12/coronavirus-takes-poultry-worker-celso-mendoza-column/3109292001/.
173. Gangitano A (2020). "Trump uses Defense Production Act to order meat processing plants to stay open." *The Hill*, 28 April. https://thehill.com/homenews/administration/495175-trump-uses-defense-production-act-to-order-meat-processing-plants-to.
174. CDC and OSHA (2020). *Meat and Poultry Processing Workers and Employers: Interim Guidance from CDC and the Occupational Safety and Health Administration (OSHA)*. 12 May. https://www.cdc.gov/coronavirus/2019-ncov/community/organizations/meat-poultry-processing-workers-employers.html; Mayer J (2020). "Back to the jungle." *The New Yorker*, 20 July. https://www.newyorker.com/magazine/2020/07/20/how-trump-is-helping-tycoons-exploit-the-pandemic.
175. Mayer J (2020). "Back to the jungle." *The New Yorker*.
176. Ibid.
177. Heer J (2020). "Meatpacking plants are a front in the Covid-19 class war." *The Nation*, 29 April. https://www.thenation.com/article/politics/meat-packing-coronavirus-class-war/; Conley J (2020). "'About as evil as it gets': As state reopens, Ohio urges employers to snitch on workers who stay home due to Covid-19 concerns." *Common Dreams*, 8 May. https://www.commondreams.org/news/2020/05/08/about-evil-it-gets-state-reopens-ohio-urges-employers-snitch-workers-who-stay-home.
178. Cockery M, D Yaffe-Bellany, and D Kravitz (2020). "As meatpacking plants reopen, data about worker illness remains elusive." *New York Times*, 25 May. https://www.nytimes.com/2020/05/25/business/coronavirus-meatpacking-plants-cases.html; Kendall L (2020). "Revealed: Covid-19 outbreaks at meat-processing plants in US being kept quiet." *The Guardian*,

1 July. https://www.theguardian.com/environment/2020/jul/01/revealed-covid-19 -outbreaks-meat-processing-plants-north-carolina.

179. Grabell M, C Perlman and B Yeung (2020). "Emails reveal chaos as meat-packing companies fought health agencies over COVID-19 outbreaks in their plants." *ProPublica*, 12 June. https://www.propublica.org/article/emails-reveal-chaos-as-meatpacking-companies-fought-health-agencies-over-covid-19-outbreaks-in-their-plants.

180. Shanker D and J Skerritt (2020). "Tyson reinstates policy that penalizes absentee workers." Bloomberg, 3 June. https://finance.yahoo.com/news/tyson-reinstates-policy-penalizes-absentee-012737585.html.

181. Mayer J (2020). "Back to the jungle."

182. Grabell M, C Perlman and B Yeung (2020). "Emails reveal chaos as meat-packing companies fought health agencies over COVID-19 outbreaks in their plants."

183. Genoways T (2014). *The Chain: Farm, Factory, and the Fate of our Food*; Freshour C (2020). "Poultry and prisons: Toward a general strike for abolition." *Monthly Review*, 1 July. https://monthlyreview.org/2020/07/01/poultry-and-prisons/.

184. Garfield L (2016). "The world's biggest meat producer is planning to test out robot butchers." *Business Insider*, 26 October. https://www.businessinsider.com/jbs-meatpacking-testing-robot-butchers-2016-10; Almeida I and J Skerritt (2020). "U.S. meat-plant changes signal end of the 99-cent chicken." Bloomberg, 12 May. https://www.bloomberg.com/news/articles/2020-05-12/human-cost-signals-end-to-99-cent-chicken-for-u-s-meat-packers.

185. Berry W (2015). "Farmland without farmers." *The Atlantic*, 19 March. https://www.theatlantic.com/national/archive/2015/03/farmland-without-farmers/388282/; Johnson KM and DT Lichter (2019). "Rural depopulation: Growth and decline processes over the past century." *Rural Sociology*, 84(1):3–27.

186. Benesh M and J Hayes (2020). "Work conditions make farmworkers uniquely vulnerable to COVID-19." *EWG News and Analysis*, 13 May. https://www.ewg.org/news-and-analysis/2020/05/work-conditions-make-farmworkers -uniquely-vulnerable-covid-19.

187. Royte E (2020). "Cases surge in America's tomato growing capital." *Mother Jones*, 5 June. https://www.motherjones.com/food/2020/06/cases-surge-in-americas -tomato-growing-capital/.

188. Dorning M and Skerritt (2020). "Every single worker has Covid at one U.S. farm on eve of harvest." Bloomberg, 29 May. https://finance.yahoo.com/news/every-single-worker-covid-one-100000688.html.

189. Ordoñez F (2020). "White House seeks to lower farmworker pay to help agriculture industry." NPR, 10 April. https://www.npr.org/2020/04/10/832076074/white-house-seeks-to-lower-farmworker-pay-to-help-agriculture-industry.

190. Horowitz R (1997). *"Negro and White, Unite and Fight!" A Social History of Industrial Unionism in Meatpacking, 1930–1990*. University of Illinois

Press, Urbana; Halpern R and R Horowitz (1999). *Meatpackers: An Oral History of Black Packinghouse Workers in Their Struggle for Racial and Economic Equality.* Monthly Review Press, New York; Johnson W (2013). *River of Dark Dreams: Slavery and Empire in the Cotton Kingdom.* Harvard University Press, Cambridge, MA; Genoways T (2014). *The Chain: Farm, Factory, and the Fate of our Food*; Rosenthal C (2018). *Accounting for Slavery: Masters and Management.* Harvard University Press, Cambridge, MA; Jackson J and R Salvador (2020). "'Our food system is very much modeled on plantation economics." FAIR, 13 May. https://fair.org/home/our-food-system-is-very-much-modeled-on-plantation-economics/. Freshour C (2020). "Poultry and prisons: Toward a general strike for abolition."

191. Scott D (2020). "Covid-19's devastating toll on black and Latino Americans, in one chart." *Vox,* 17 April. https://www.vox.com/2020/4/17/21225610/us-coronavirus-death-rates-blacks-latinos-whites.

192. Gennetian LA and MS Johnson (2020). "Work-based risks to Latino workers and their families from COVID-19." *Econofact,* 26 May. https://econofact.org/work-based-risks-to-latino-workers-and-their-families-from-covid-19.

193. Benton A (2020). "Race, epidemics, and the viral economy of health expertise." *The New Humanitarian,* 4 February. https://www.thenewhumanitarian.org/opinion/2020/02/04/Coronavirus-xenophobia-outbreaks-epidemics-social-media; Frias L (2020). "A Wisconsin chief justice faced backlash for blaming a county's coronavirus outbreak on meatpacking employees, not 'regular folks'." *Business Insider,* 7 May. https://www.businessinsider.com/chief-justice-condemned-for-blaming-coronavirus-spread-on-meatpackers-2020-5.

194. Philpott T (2020). "Republicans keep blaming workers for coronavirus outbreaks at meat plants." *Mother Jones,* 8 May. https://www.motherjones.com/food/2020/05/republicans-keep-blaming-workers-for-coronavirus-outbreaks-at-meat-plants/.

195. Pitt D (2020). "Worker advocates file meat plants discrimination complaint." AP News, 9 July. https://apnews.com/41f90b02d3eeedfc9035f4748f46ab3c.

196. Pitt D (2020). "CDC: Minorities affected much more in meatpacking outbreaks." AP News, 8 July. https://apnews.com/12c6f7dd8888b7f2a174ae4ba8f06b67.

197. Freshour C and B Williams (2020). "Abolition in the time of COVID-19." *Antipode Online,* 9 April. https://antipodeonline.org/2020/04/09/abolition-in-the-time-of-covid-19/; Freshour C (2020). "Poultry and prisons: Toward a general strike for abolition."

198. Eason JM (2017). *Big House on the Prairie: Rise of the Rural Ghetto and Prison Proliferation.* University of Chicago Press, Chicago.

199. Flagg A and J Neff (2020). "Why jails are so important in the fight against coronavirus." *New York Times,* 31 March. https://www.nytimes.com/2020/03/31/upshot/coronavirus-jails-prisons.html.

200. Yang J, Y Zheng, X Gou, K Pu, Z Chen, et al. (2020). "Prevalence of comorbidities and its effects in patients infected with SARS-CoV-2: a systematic review and meta-analysis." *International Journal of Infectious Diseases* 94:91-95; Baker MG, TK Peckham, and NS Seixas (2020). "Estimating the burden of United States workers exposed to infection or disease: A key factor in containing risk of COVID-19 infection." *PLoS ONE 15*(4): e0232452.

201. Wu X, RC Nethery, BM Sabath, D Braun, and F Dominici (2020). "Exposure to air pollution and COVID-19 mortality in the United States: A nationwide cross-sectional study." medRxiv, 27 April. https://www.medRxiv.org/content/10.1101/2020.04.05.20054502v2; Friedman L (2020). "New research links air pollution to higher coronavirus death rates." *New York Times,* 17 April. https://www.nytimes.com/2020/04/07/climate/air-pollution-coronavirus-covid.html.

202. Tessum CW, JS Apte, AL Goodkind, NZ Muller, KA Mullins, et al. (2019). "Inequity in consumption of goods and services adds to racial–ethnic disparities in air pollution exposure." *PNAS,* 116 (13): 6001–6006.

203. Schulte F, E Lucas, J Rau, L Szabo, and J Hancock (2020). "Millions of older Americans live in counties with no ICU beds as pandemic intensifies." *Kaiser Health News,* 20 March. https://khn.org/news/as-coronavirus-spreads-widely-millions-of-older-americans-live-in-counties-with-no-icu-beds/.

204. North Dakota Department of Health. "Map of ventilators by county." https://www.health.nd.gov/map-ventilators-county.

205. Knapp EA, U Bilal, LT Dean, M Lazo, and DD Celentano (2019). "Economic insecurity and deaths of despair in US counties." *American Journal of Epidemiology,* 188(12):2131-2139; Dobis EA, HM Stephens, M Skidmore, and SJ Goetz (2020). "Explaining the spatial variation in American life expectancy." *Social Science & Medicine*, 246:112759.

206. Frydl K (2016). "The Oxy electorate: A scourge of addiction and death siloed in flyover country." *Medium,* 16 November. https://medium.com/@kfrydl/the-oxyelectorate-3fa62765f837; Bor J (2017). "Diverging life expectancies and voting patterns in the 2016 US presidential election." *American Journal of Public Health* 107(10):1560–1562; Wallace D and R Wallace (2019). *Politics, Hierarchy, and Public Health: Voting Patterns in the 2016 US Presidential Election.* Routledge, New York.

207. Rynard P (2020). "Louisa County COVID-19 rate now worse than New York State." *Iowa Starting Line,* 13 April. https://iowastartingline.com/2020/04/13/louisa-county-covid-19-rate-now-worse-than-new-york-state/.

208. Foley RJ (2020). "Outbreak at Iowa pork plant was larger than state reported." AP News, 22 July. https://apnews.com/85a02d9296053980ea47eba97f920707.

209. Hudson B (2020). "Coronavirus in Minnesota: Nobles County reaches 866 known cases of COVID-19, most traced to JBS pork plant." CBS Minnesota, 1 May. https://minnesota.cbslocal.com/2020/05/01/coronavirus-in-minnesota-nobles-county-reaches-866-known-cases-of-

covid-19-most-traced-to-jbs-pork-plant/; Butte Lab (2020). "Cumulative cases by Minnesota county—linear plot: Nobles County." COVID-19 County Tracker, 8 June. https://comphealth.ucsf.edu/app/buttelabcovid.

210. Belz A, E Flores and G Stanley (2020). "As coronavirus loomed, Worthington pork plant refused to slow down." *Star Tribune*, 16 May. https://www.startribune.com/as-coronavirus-loomed-worthington-pork-plant-refused-to-slow-down/570516612/.

211. MovieClips (2015). *"Fargo* (1996)—The Wood Chipper Scene (11/12) | Movieclips." https://www.youtube.com/watch?v=0YzsWVUO-_o; Forum News Service (2020). "Wood chippers employed to help compost thousands of excess hogs near Worthington plant." *Pioneer Press*, 6 May. https://www.twincities.com/2020/05/06/wood-chippers-employed-to-help-compost-thousands-of-excess-hogs-near-worthington-plant/; Corkery M and D Yaffe-Bellany (2020). "Meat plant closures mean pigs are gassed or shot instead." *New York Times*, 14 May. https://www.nytimes.com/2020/05/14/business/coronavirus-farmers-killing-pigs.html.

212. Greenwald G (2020). "Hidden video and whistleblower reveal gruesome mass-extermination method for Iowa pigs amid pandemic." *The Intercept*, 29 May. https://theintercept.com/2020/05/29/pigs-factory-farms-ventilation-shutdown-coronavirus/.

213. Van Pykeren S (2020). "These photos show the staggering food bank lines across America." *Mother Jones*, 13 April. https://www.mother jones.com/food/2020/04/these-photos-show-the-staggering-food-bank-lines-across-america/; Ransom E, EM DuPuis, and MR Worosz (2020). "Why farmers are dumping milk down the drain and letting produce rot in fields." *The Conversation*, 23 April. https://theconversation.com/why-farmers-are-dumping-milk-down-the-drain-and-letting-pro-duce-rot-in-fields-136567; Jackson J and R Salvador (2020). "'Our food system is very much modeled on plantation economics'"; Nylen L and L Crampton (2020). "'Something isn't right': U.S. probes soaring beef prices." *Politico*, 25 May. https://www.politico.com/news/2020/05/25/meatpackers-prices-coronavirus-antitrust-275093.

214. Housing Assistance Council (2020). *Update: COVID-19 in Rural America —May 29, 2020.* http://www.ruralhome.org/whats-new/mn-coronavirus/1819-covid-19-in-rural-america-update.

215. Fisher A (2017) *The Unholy Alliance between Corporate America and Anti-Hunger Groups.* MIT Press, Cambridge, MA; Glauber JW, DA Sumner, and PE Wilde (2017). *Poverty, Hunger, and US Agricultural Policy: Do Farm Programs Affect the Nutrition of Poor Americans?* AEI Paper & Studies; Lloyd JL (2019). "From farms to food deserts: Food insecurity and older rural Americans." *Generations*, 43(2):24–32.

216. Wallace RG (2016). "Made in Minnesota." In *Big Farms Make Big Flu: Dispatches on Infectious Disease, Agribusiness, and the Nature of Science.* Monthly Review Press, New York, pp 347–358.

217. Cockery M and D Yaffe-Bellany (2020). "As meat plants stayed open to feed Americans, exports to China surged." *New York Times*, 16 June. https://www.nytimes.com/2020/06/16/business/meat-industry-china-pork.html.
218. Ibid.
219. Practical Farmers of Iowa (2020). *Strategies for Strange Times*. One-hour virtual farmer meet-up. https://practicalfarmers.org/events/covid-19-changes-and-challenges-virtual-farmer-meet-up/.
220. Westervelt E (2020). "As food supply chain breaks down, farm-to-door CSAs take off." NPR, 10 May. https://www.npr.org/2020/05/10/852512047/as-food-supply-chain-breaks-down-farm-to-door-csas-take-off.
221. Niche Meat Processor Assistance Network (2020). "Mobile slaughter and processing." https://www.nichemeatprocessing.org/mobile-unit-overview/.
222. Map (2020). "Open Food Network USA." https://openfoodnetwork.net/map; Hill P (2020). "Mozilla picks recipients for its COVID-19 Solutions Fund." *Neowin*, 9 June. https://www.neowin.net/news/mozilla-picks-recipients-for-its-covid-19-solutions-fund.
223. Bruce AB and RLS Castellano (2017). "Labor and alternative food networks: challenges for farmers and consumers." *Renewable Agriculture and Food Systems* 32(5): 403-416; Chappell MJ (2018). *Beginning to End Hunger: Food and the Environment in Belo Horizonte, Brazil and Beyond*. University of California Press, Oakland; Motzer N (2019). "'Broad but not deep': regional food hubs and rural development in the United States." *Social & Cultural Geography*, 20(8): 1138–1159.

Square Roots

224. Monk R (1990). *Ludwig Wittgenstein: The Duty of Genius*. Penguin Books, New York.
225. Wallace RG. "Notes on a novel coronavirus." This volume.
226. Fearnley L (2013). "The birds of Poyang Lake: Sentinels at the interface of wild and domestic." *Limn* 3. https://limn.it/issues/sentinel-devices/.
227. Banaji J (1990). "Illusions about the peasantry: Karl Kautsky and the Agrarian Question." *The Journal of Peasant Studies*, 17 (2): 288–230.
228. Miller A (2020). "COVID-19 sees focus on periurban agriculture, food security." *Stock & Land*, 6 April. https://www.stockandland.com.au/story/6713432/coronavirus-sharpens-focus-on-food-security/.
229. Wallace RG (2019). "Redwashing capital: Left tech bros are honing Marx into a capitalist tool." *Uneven Earth*, 11 July. http://unevenearth.org/2019/07/redwashing-capital/.
230. AmericanExperimentMN (2018). "Support mining in Minnesota." YouTube, 29 August. https://www.youtube.com/watch?v=VHGPY4sssqc&feature=youtube.

Midvinter-19

231. AwesomeMovieScenesHD (2019). "Ending and the cult ritual/*Midsommar*

(2019) Movie Clip HD." YouTube, 26 September. https://www.youtube. com/watch?v=InRMXiwFPxE.

232. Mizkex (2019). "*Midsommar* Ending Without Music (DISTURBING)." YouTube, 13 December. https://www.youtube.com/watch?v=qpiLMuwReyA.

233. Beaumont P (2020). "Where did Covid-19 come from? What we know about its origins." *The Guardian*, 1 May. https://www.theguardian.com/ world/2020/may/01/could-covid-19-be-manmade-what-we-know-about-origins-trump-chinese-lab-coronavirus; Cummins R (2020). "Murder most foul: the perps behind COVID-19." Organic Consumers Association website, 29 April. https://www.organicconsumers.org/blog/ murder-most-foul-perps-behind-covid-19; Musto J (2020). "Tom Cotton calls on China to produce evidence that disputes Wuhan lab as source of COVID-19." Fox News, 5 May. https://www.foxnews.com/media/ tom-cotton-calls-on-china-to-produce-evidence-disputing-lab-theory.

234. Wallace RG, et al. (2015). "The dawn of Structural One Health: a new science tracking disease emergence along circuits of capital." *Social Science and Medicine* 129: 68–77.

235. Wallace RG, A Liebman, LF Chaves, and R Wallace (2020). "COVID-19 and circuits of capital." This volume.

236. Schneider M (2017). "Wasting the rural: Meat, manure, and the politics of agro-industrialization in contemporary China." *Geoforum* 78: 89–97.

237. Wallace R, et al. (2018). *Clear-Cutting Disease Control: Capital-Led Deforestation, Public Health Austerity, and Vector-Borne Infection*. Springer International Publishing, Cham.

238. Wallace R and RG Wallace (eds), *Neoliberal Ebola: Modeling Disease Emergence from Finance to Forest and Farm*. Springer International Publishing, Cham.

239. Wallace RG (2016). *Big Farms Make Big Flu*. Monthly Review Press, New York.

240. Tang XC, et al. (2006). "Prevalence and genetic diversity of coronaviruses in bats from China." *Journal of Virology* 80(15): 7481–7490; Hu W, et al. (2005). "Development and evaluation of a multitarget real-time Taqman reverse transcription-PCR assay for detection of the severe acute respiratory syndrome-associated coronavirus and surveillance for an apparently related coronavirus found in masked palm civets." *Journal of Clinical Microbiology* 43(5): 2041–2046.

241. Wu Z, et al. (2016). "Deciphering the bat virome catalog to better understand the ecological diversity of bat viruses and the bat origin of emerging infectious diseases." *The ISME Journal* 10(3): 609-620; Fan Y, K Zhao, Z-L Shi, and P Zhou (2019). "Bat coronaviruses in China." *Viruses*, 11(3): 210.

242. Li X, Y Gao, C Wang, and B Sun (2020). "Influencing factors of express delivery industry on safe consumption of wild dynamic foods." *Revista Científica*. 30(1): 393–403.

243. Forster P, L Forster, C Renfrew, and M Forster (2020). "Phylogenetic

analysis of SARS-CoV-2 genomes." *Proceedings of the National Academy of the Sciences* 117(17): 9241–9243.

244. Ridley M (2003). *Evolution*, Third Edition. Wiley-Blackwell, Hoboken, NJ.

245. Pipes L, H Wang, JP Huelsenbeck, and R Nielsen (2020). "Assessing uncertainty in the rooting of theSARS-CoV-2 phylogeny." bioRxiv, 20 June. https://www.biorxiv.org/content/10.1101/2020.06.19.160630v1.full.pdf.

246. Zhou H et al (2020). "A novel bat coronavirus reveals natural insertions at the S1/S2 cleavage site of the Spike protein and a possible recombinant origin of HCov-19." bioRxiv, 11 March. https://www.bioRxiv.org/content/10.1101/2020.03.02.974139v3.full.

247. Xiao K, et al. (2020). "Isolation and characterization of 2019-nCoV-like coronavirus from Malayan pangolins." bioRxiv, 20 February. https://www.bioRxiv.org/content/10.1101/2020.02.17.951335v1.

248. Schneider M (2016). "Dragon head enterprises and the state of agribusiness in China." *Journal of Agrarian Change* 17(1): 3–21; Bharucha Z and J Pretty (2010). "The roles and values of wild foods in agricultural systems." *Philosophical Transactions of the Royal Society B* 365: 2913–2926.

249. Challender D (2019). "Evaluating the feasibility of pangolin farming and its potential conservation impact." *Global Ecology and Conservation* 20: 00714; Nash HC and C Waterman (2019). *Pangolins: Science, Society, and Conservation*. Elsevier, Amsterdam.

250. Standaert M (2020). "Coronavirus closures reveal vast scale of China's secretive wildlife farm industry." *The Guardian*, 24 February. https://www.theguardian.com/environment/2020/feb/25/coronavirus-closures-reveal-vast-scale-of-chinas-secretive-wildlife-farm-industry.

251. Yan W (2019). "The plight of the pangolin in China." *chinadialogue*, 6 May. https://www.chinadialogue.net/article/show/single/en/11275-The-plight-of-the-pangolin-in-China.

252. Liu P, W Chen, and JP Chen. "Viral metagenomics revealed sendai virus and coronavirus infection of Malayan pangolins (*Manis javanica*)." *Viruses* 11(11): 979.

253. Olson S, et al. (2014). "Sampling strategies and biodiversity of influenza A subtypes in wild birds." *PLoS ONE*, 5 March. https://journals.plos.org/plosone/article?id=10.1371/journal.pone.0090826; Gossner C (2014). "Human-dromedary camel interactions and the risk of acquiring zoonotic Middle East Respiratory Syndrome coronavirus syndrome." *Zoonoses and Public Health* 63(1): 1–9.

254. Nijman V, MX Zhang, and CR Shepherd (2016). "Pangolin trade in the Mong La wildlife market and the role of smuggling pangolins into China." *Global Ecology and Conservation* 5: 118–126.

255. Ling X, J Guan, W Lau, and Y Xiao (2016). "An overview of pangolin trade in China." *Traffic Briefing*, September 2016. Available online at https://d2ouvy59p0dg6k.cloudfront.net/downloads/briefing_paper_of_pangolin_trade.pdf.

256. Neme, L (2016). "Myanmar feeds China's pangolin appetite." *National Geographic*, 19 January. https://www.nationalgeographic.com/news/2016/01/160119-pangolins-china-myanmar-wildlife-trafficking/.

257. Lacroix A (2017). "Genetic diversity of coronaviruses in bats in Lao PDR and Cambodia." *Infection, Genetics, and Evolution* 48: 10-18; Zhou P, et al. (2018). "Fatal swine acute diarrhoea syndrome caused by an HKU2-related coronavirus of bat origin." *Nature* 556(7700): 255–258.

258. Zhao Y, K Zhang, Y Fu, and H Zhang (2012). "Examining land-use/land-cover change in the Lake Dianchi watershed of the Yunnan-Guizhou Plateau of Southwest China with remote sensing and GIS techniques: 1974–2008." *International Journal of Environmental Research and Public Health* 9(11): 3843-3865.

259. Wallace RG (2008). "Review of Paul Torrence's *Combating the Threat of Pandemic Influenza: Drug Discovery Approaches*." *The Quarterly Review of Biology* 83(3): 327–328.

260. Adam D, D Magee, C Bui, M Scotch, and C MacIntyre (2017). "Does influenza pandemic preparedness and mitigation require gain-of-function research?" *Influenza and Other Respiratory Diseases* 11(4): 306–310.

261. Forst CV, et al. (2017). "Integrative gene network analysis identifies key signatures, intrinsic networks and host factors for influenza virus A infections." *Systems Biology and Applications* 3: 35.

262. Husseini, S (2020). "Did this virus come from a lab? Maybe not — but it exposes the danger of a biowarfare arms race." *Salon*, 24 April https://www.salon.com/2020/04/24/did-this-virus-come-from-a-lab-maybe-not--but-it-exposes-the-threat-of-a-biowarfare-arms-race; Guterl, F (2020). "Dr. Fauci backed controversial Wuhan lab with U.S. dollars for risky coronavirus research." *Newsweek*, 28 April. https://www.newsweek.com/dr-fauci-backed-controversial-wuhan-lab-millions-us-dollars-risky-coronavirus-research-1500741.

263. Wallace RG (2016). *Big Farms Make Big Flu.*

264. Okamoto K, A Liebman, and RG Wallace (2019). "At what geographic scales does agricultural alienation amplify foodborne disease outbreaks? A statistical test for 25 U.S. states, 1970–2000." medRxiv, 8 January. https://www.medRxiv.org/content/10.1101/2019.12.13.19014910v2.

265. Wallace RG, et al. (2015). "The dawn of Structural One Health: a new science tracking disease emergence along circuits of capital."

266. Li HY, et al. (2020). "A qualitative study of zoonotic risk factors among rural communities in southern China." *Int Health* 12(2): 77–85.

267. Menachery VD, et al. (2016). "A SARS-like cluster of circulating bat coronaviruses shows potential for human emergence." *Nature Medicine* 21(12): 1508–1513.

268. Ren W, et al. (2008). "Difference in receptor usage between severe acute respiratory syndrome (SARS) coronavirus and SARS-like coronavirus of bat origin." *Journal of Virology* 82(4): 1899–1907; Li W, et al. (2005). "Bats

are natural reservoirs of SARS-like coronaviruses." *Science* 310(5748): 676–679.

269. Hu B, et al. (2017). "Discovery of a rich gene pool of bat SARS-related coronaviruses provides new insights into the origin of SARS coronavirus." *PLoS Pathogens.* 13(11): e1006698; Democracy Now (2020). "'Pure baloney': Zoologist debunks Trump's COVID-19 origin theory, explains animal-human transmission." *Democracy Now,* 16 April. https://www.democracynow.org/2020/4/16/peter_daszak_coronavirus.

270. Zhou P, et al. (2018). "Fatal swine acute diarrhoea syndrome caused by an HKU2-related coronavirus of bat origin." *Nature* 556(7700): 255–258.

271. Wallace RG (2011). "A dangerous method." *Farming Pathogens* blog, 28 December. https://farmingpathogens.wordpress.com/2011/12/28/a-dangerous -method/.

272. Branswell H (2015). "SARS-like virus in bats shows potential to infect humans, study finds." *STAT,* 9 November https://www.statnews.com/2015/11/09/sars-like-virus-bats-shows-potential-infect-humans-study-finds/.

273. BBC News (2020). "Coronavirus: Trump stands by China lab origin theory for virus." *BBC News,* 1 May. https://www.bbc.com/news/world-us -canada-52496098.

274. Zarley BD (2020). "Genetic evidence debunks coronavirus conspiracy theories, scientists say." *freethink,* 5 April. https://www.freethink.com/articles/coronavirus-conspiracy-theories.

275. Garrett L (2011). "The bioterrorist next door." *Foreign Policy,* 15 December. https://foreignpolicy.com/2011/12/15/the-bioterrorist-next-door/#sthash.FbHXLDbC.dpbs.

276. Boeckel TP (2013). "The Nosoi commute: a spatial perspective on the rise of BSL-4 laboratories in cities." *arXiv* 1312.3283. https://arxiv.org/abs/1312.3283.

277. Wallace RG (2013). "Homeland." *Farming Pathogens* blog, 16 December. https://farmingpathogens.wordpress.com/2013/12/16/homeland/.

278. Furmanski, M (2014). "Threatened pandemics and laboratory escapes: Self-fulfilling prophecies." *Bulletin of the Atomic Scientists,* 31 March. https://thebulletin.org/2014/03/threatened-pandemics-and-laboratory -escapes-self-fulfilling-prophecies/.

279. Zhang H, et al. (2018). "Evaluation of MICRO-CHEM PLUS as a disinfectant for Biosafety Level 4 laboratory in China." *Applied Biosafety* 23(1): 32–38.

280. Anonymous (2020). "Evidence SARS-CoV-2 emerged from a biological laboratory in Wuhan, China." 2 May. https://project-evidence.github.io/#%28part._authors%29.

281. Daszak P (2014). "Understanding the risk of bat coronavirus emergence." *Grantome.com* documentation of grant funded by the National Institute of Health. https://grantome.com/grant/NIH/R01-AI110964-06#panel-funding.

282. Owermohle S (2020). "Trump cuts research on bat-human virus

transmission over China ties." *Politico*, 27 April. https://www.politico.com/ews/2020/04/27/trump-cuts-research-bat-human-virus-china-213076.

283. Wallace RG, A Liebman, LF Chaves, and R Wallace (2020). "COVID-19 and circuits of capital."

284. Piper K (2020). "Why some labs work on making viruses deadliner—and why they should stop." *Vox*, 1 May. https://www.vox.com/2020/5/1/21243148/why-some-labs-work-on-making-viruses-deadlier-and-why-they-should-stop.

285. Kirchgaessner S, E Graham-Harrison, and L Kuo (2020). "China clamping down on coronavirus research, deleted pages suggest." *The Guardian*, 11 April. https://www.theguardian.com/world/2020/apr/11/china-clamping-down-on-coronavirus-research-deleted-pages-suggest.

286. Badiou A (2005). *Metapolitics*. Verso, London.

287. Ali T (2018). *The Extreme Center: A Second Warning*. Verso, London.

Blood Machines

288. Fox A (2020). "The race for a coronavirus vaccine runs on horseshoe crab blood." *Smithsonian Magazine*, 8 June. https://www.smithsonianmag.com/smart-news/race-coronavirus-vaccine-runs-horseshoe-crab-blood-180975048/.

289. Ding JL and B Ho (2010). "Endotoxin detection--from *Limulus* amebocyte lysate to recombinant Factor C." In X Wang and PJ Quinn (eds), *Endotoxins: Structure, Function and Recognition*. Springer, Cham, pp 187–208.

290. Fortey R (2012) *Horseshoe Crabs and Velvet Worms: The Story of the Animals and Plants That Time Has Left Behind*. Knopf, New York.

291. Morton T (2016). *Dark Ecology: For a Logic of Future Existence*. Columbia University Press, New York; Badiou A (2019). *I Know There Are So Many of You*. Translated by S Spitzer. Polity Press, Cambridge, UK.

292. Chesler C (2017). "Why biomedical companies are bleeding horseshoe crabs to the brink of extinction." *Esquire*, 14 April. https://www.esquire.com/news-politics/a54516/blood-of-horseshoe-crab/.

293. Ibid.

294. Ding JL and B Ho (2010). "Endotoxin detection—from Limulus Amebocyte Lysate to Recombinant Factor C."

295. Gorman J (2020). "Tests for coronavirus vaccine need this ingredient: horseshoe crabs." *New York Times*, 3 June. https://www.nytimes.com/2020/06/03/science/coronavirus-vaccine-horseshoe-crabs.html.

296. Miller J (2020). "Drugs standards group nixes plan to kick pharma's crab blood habit." Reuters, 30 May. https://www.reuters.com/article/us-lonza-crabs/drugs-standards-group-nixes-plan-to-kick-pharmas-crab-blood-habit-idUSKBN2360MB.

297. Gorman J (2020). "Tests for coronavirus vaccine need this ingredient: horseshoe crabs."

298. Ajl M and RG Wallace. "The bright bulbs." This volume.

299. Wallace RG (2008). "Book review: 'Combating the Threat of Pandemic Influenza: Drug Discovery Approaches.'" *Quarterly Review of Biology* 83:327–328

300. Chesler C (2017). "Why biomedical companies are bleeding horseshoe crabs to the brink of extinction."
301. Moore JW (2015). *Capitalism in the Web of Life: Ecology and the Accumulation of Capital.* Verso, New York.

The Origins of Industrial Agricultural Pathogens

302. Broswimmer FJ (2002). *Ecocide: A Short History of the Mass Extinction of Species.* Pluto Press, London; Dawson A (2016) *Extinction: A Radical History.* OR Books, New York.
303. Foley J, N Ramankutty, KA Brauman, ES Cassidy, JS Gerber, et al. (2011). "Solutions for a cultivated planet." *Nature* 478: 337–342; Valladares G, L Cagnolo, and A Salvo (2012). "Forest fragmentation leads to food web contraction." *Oikos* 121(2): 299–305; Sinclair ARE and A Dobson (2015). "Conservation in a human-dominated world." In ARE Sinclair, KL Metzger, SAR Mduma, and JM Fryxell (eds), *Serengeti IV: Sustaining Biodiversity in a Coupled Human-Natural System.* University of Chicago Press, Chicago, IL, pp 1-10; Ferreira S, F Martínez-Freiría, J-P Boudot, M El Haissoufi, and N Bennas (2015). "Local extinctions and range contraction of the endangered *Coenagrion mercuriale* in North Africa." *International Journal of Odonatology* 18(2): 137–152; Wolf C and WJ Ripple (2017). "Range contractions of the world's large carnivores." *Royal Society Open Science* 4: 170052.
304. Crowl TA, TO Crist, RR Parmenter, G Belovsky, and AE Lugo (2008). "The spread of invasive species and infectious disease as drivers of ecosystem change." *Frontiers in Ecology and the Environment* 6(5): 238–246; Paini DR, AW Sheppard, DC Cook, PJ De Barro, SP Worner, and MB Thomas (2016). "Global threat to agriculture from invasive species." *PNAS* 113(27):7575–7579; Chapman D, BV Purse, HE Roy, and JM Bullock (2017). "Global trade networks determine the distribution of invasive non-native species." *Global Ecology and Biogeography* 26(8): 907–917; Wyckhuys KAG, AC Hughes, C Buamas, AC Johnson, L Vasseur, L Reymondin, J-P Deguine, and S Sheil (2019). "Biological control of an agricultural pest protects tropical forests." *Commun Biol.* 2: 10.
305. Hoffmann I (2010). "Livestock biodiversity." *Rev Sci Tech.* 29(1): 73–86.
306. Bast F (2016). "Primary succession recapitulates phylogeny." *J Phylogenetics Evol Biol* 4: 1.
307. Lefebvre H (1974; 2000). *The Production of Space.* Blackwell Publishers, Oxford; Foster JB (1999). "Marx's theory of metabolic rift: Classical foundations for environmental sociology." *American Journal of Sociology* 105(2): 366–405; Malm, A (2016) *Fossil Capital: The Rise of Steam Power and the Roots of Global Warming.* Verso, New York; Okamoto, KW, A Liebman, and RG Wallace (2020) "At what geographic scales does agricultural alienation amplify foodborne disease outbreaks? A statistical test for 25 U.S. States, 1970–2000." medRxiv, 8 January. https://www.medRxiv.org/content/10.110 1/2019.12.13.19014910v2; Wallace RG, K Okamoto, and A Liebman (2020)

"Gated ecologies." In DB Monk and M Sorkin (eds), *Between Catastrophe and Redemption: Essays in Honor of Mike Davis*. Terreform/OR Books, New York.

308. Illich I (1971; 2000) *Deschooling Society*. Marion Boyars Publishers, London; Illich I (2013). *Beyond Economics and Ecology: The Radical Thought of Ivan Ilich*. Marion Boyars Publishers, London; Ehgartner U, P Gould, and M Hudson (2017). "On the obsolescence of human beings in sustainable development." *Global Discourse* 7(1): 66–83; Galluzzo A (2018). "The singularity in the 1790s: Toward a prehistory of the present with William Godwin and Thomas Malthus." *B20*, 17 September. https://www.boundary2.org/2018/09/galluzzo/.

309. Gignoux CR, BM Henn, and JL Mountain (2011). "Rapid, global demographic expansions after the origins of agriculture." *PNAS* 108(15): 6044–6049; Bocquet-Appel J-P (2011). "The agricultural demographic transition during and after the agriculture inventions." *Current Anthropology* 52(S4): S497-S510; Wallace RG and R Kock (2012). "Whose food footprint? Capitalism, agriculture and the environment." *Human Geography* 5(1): 63–83; Gavin MC, PH Kavanagh, HJ Haynie, C Bowern, CR Ember, et al. (2018). "The global geography of human subsistence." *Royal Society Open Science*, 5(9): 171897.

310. Chappell MJ (2018). *Beginning to End Hunger: Food and the Environment in Belo Horizonte, Brazil, and Beyond*. University of California Press, Berkeley; Kallis G (2019). *Limits: Why Malthus Was Wrong and Why Environmentalists Should Care*. Stanford University Press, Stanford, CA.

311. Stuart D and R Gunderson (2018). "Nonhuman animals as fictitious commodities: Exploitation and consequences in industrial agriculture." *Society & Animals* 1 (aop): 1–20; Miles C (2019). "The combine will tell the truth: On precision agriculture and algorithmic rationality." *Big Data & Society* January-June: 1–12. https://doi.org/10.1177/2053951719849444; Wurgaft, B.A. 2019. *Meat Planet: Artificial Flesh and the Future of Food*. University of California Press, Berkeley, CA; Wallace RG (2019). "From agribusiness to agroecology: Escaping the market of Dr. Moreau." 10 November. Session on "Utopia, degrowth, and ecosocialism", Historical Materialism, London. https://www.youtube.com/watch?v=ws8CsVJnnc0.

312. Levins R and JH Vandermeer (1990). "The agroecosystem embedded in a complex ecological community." In CR Carroll, JH Vandermeer, and PM Rosset (eds), *Agroecology*. McGraw-Hill, New York, pp 341–362; Rosset PM and ME Martinez-Torres (2013). "La Via Campesina and agroecology." In *La Via Campesina's Open Book: Celebrating 20 Years of Struggle and Hope*. http://nyeleni.pl/wp-content/uploads/2018/01/La-Via-Campesina-and -Agroecology.pdf; Perfecto I, J Vandermeer, and A Wright (2019). *Nature's Matrix: Linking Agriculture, Biodiversity Conservation and Food Sovereignty*. 2nd Edition. Earthscan, London; Wallace RG, K Okamoto, and A Liebman (2020) "Gated ecologies."

313. Alders R, M Nunn, B Bagnol, J Cribb, R Kock, and J Rushton (2016). "Approaches to fixing broken food systems." In M. Eggersdorfer, et al. (eds), ' *Good Nutrition: Perspectives for the 21st Century*. Krager, Basel, Switzerland, pp 132–144; Chappell MJ (2018). *Beginning to End Hunger: Food and the Environment in Belo Horizonte, Brazil, and Beyond.*

314. Amin S (1988; 2009). *Eurocentrism: Modernity, Religion, and Democracy.* Monthly Review Press, New York; Tsing A (2009). "Supply chains and the human condition." *Rethinking Marxism*, 21(2): 148–176; Coulthard GS (2014). *Red Skin, White Masks: Rejecting the Colonial Politics of Recognition.* University of Minnesota Press, Minneapolis; Vergara-Camus L (2014). *Land and Freedom: The MST, the Zapatistas and Peasant Alternatives to Neoliberalism.* Zed Books, London; Chappell MJ (2018). *Beginning to End Hunger: Food and the Environment in Belo Horizonte, Brazil, and Beyond;* Giraldo OF (2019). *Political Ecology of Agriculture: Agroecology and Post-Development.* Springer, Cham.

315. Foley J, R Defries, GP Asner, C Barford, G Bonan, SR Carpenter, et al. (2005). "Global consequences of land use." *Science* 309: 570–574; Ellis EC and N Ramankutty (2008). "Putting people in the map: anthropogenic biomes of the world." *Frontiers in Ecology and the Environment* 6(8): 439–447; Alexandratos N and J Bruinsma (2012). *World Agriculture Towards 2030/2050: The 2012 Revision. ESA Working Paper 12-03.* Agricultural Development Economics Division, Food and Agriculture Organization. http://www.fao.org/fileadmin/templates/esa/Global_persepctives/world_ag_2030_50_2012_rev.pdf; FAO (2013). *FAO Statistical Yearbook 2013.* Food and Agriculture Organization, United Nations, Rome; Wallace R, L Bergmann, L Hogerwerf, R Kock and RG Wallace (2016). "Ebola in the hog sector: Modeling pandemic emergence in commodity livestock." In RG Wallace and R Wallace (eds), *Neoliberal Ebola: Modeling Disease Emergence from Finance to Forest and Farm.* Springer, Cham; Ramankutty N, Z Mehrabi, K Waha, L Jarvis, C Kremen, M Herrero, and LH Rieseberg (2018). "Trends in global agricultural land use: Implications for environmental health and food security." *Annu Rev Plant Biol.*, 69: 789–815.

316. Goldewijk KK, A Beusen, J Doelman, and E Stehfest (2017). "Anthropogenic land use estimates for the Holocene–HYDE 3.2." *Earth Syst. Sci. Data*, 9(2): 927–953.

317. Smil V (2002). "Eating meat: Evolution, patterns, and consequences." *Population and Development Review*, 28:599–639; Van Boeckel TP, W Thanapongtharm, T Robinson, L D'Aietti, and M Gilbert (2012). "Predicting the distribution of intensive poultry farming in Thailand." *Agriculture, Ecosystems & Environment*, 149:144–153; Robinson TP, GRW Wint, G Conchedda, TP Van Boeckel, V Ercoli, E Palamara, G Cinardi, L D'Aietti, SI Hay, and M Gilbert (2014). "Mapping the global distribution of livestock." *PLoS ONE*, 9(5): e96084; Nicolas G, TP Robinson, GRW Wint, G Conchedda, G Cinardi, and M Gilbert (2016). "Using Random Forest to

improve the downscaling of global livestock census data." *PLoS ONE,* 11(3): e0150424.

318. Wallace R, L Bergmann, L Hogerwerf, R Kock, and RG Wallace (2016). "Ebola in the hog sector: Modeling pandemic emergence in commodity livestock"; FAO (2018). "FAOSTAT: Live animals." Food and Agriculture Organization of the United Nations. http://www.fao.org/faostat/en/#data/QA/visualize.

319. Steinfeld H, P Gerber, T Wassenaar, V Castel, M Rosales, and C Haan (2006). *Livestock's Long Shadow: Environmental Issues and Options.* Food and Agriculture Organization, Rome; Herrero M, P Havlík, H Valin, A Notenbaert, MC Rufino, et al. (2013). "Biomass use, production, feed efficiencies, and greenhouse gas emissions from global livestock systems." *PNAS* 110(52): 20888–20893; IPCC (2019).*Climate Change and Land.* https://www.ipcc.ch/site/assets/uploads/2019/08/Fullreport-1.pdf.

320. Global Carbon Project (2011). "Global carbon budget archive." https://www.globalcarbonproject.org/carbonbudget/archive.htm; Gerber PJ, H Steinfeld, B Henderson, A Mottet, C Opio, et al. (2013). *Tackling Climate Change Through Livestock: A Global Assessment of Emissions and Mitigation Opportunities.* FAO, Rome; Rojas-Downing MM, AP Nejadhashemi, T Harrigan, and SA Woznicki (2017). "Climate change and livestock: Impacts, adaptation, and mitigation." *Climate Risk Management* 16: 145–163.

321. FAO (2019). "GLEAM 2.0—Assessment of greenhouse gas emissions and mitigation potential." http://www.fao.org/gleam/results/en/.

322. Mészáros I (1970). *Marx's Theory of Alienation.* Merlin Press, UK; Ellis EC, PJ Richerson, A Mesoudi, J-C Svenning, J Odling-Smee, and WR Burnside (2016). "Evolving the human niche." *PNAS,* 113(31): E4436; Piperno, DR, AJ Ranere, R Dickau, and F Aceituno (2017). "Niche construction and optimal foraging theory in Neotropical agricultural origins: A re-evaluation in consideration of the empirical evidence." *Journal of Archaeological Science* 78: 214–220; Arroyo-Kalin, M, CJ Stevens, D Wengrow, DQ Fuller, and M Wollstonecroft (2017). "Civilisation and human niche construction." *Archaeology International* 20: 106–109; Levis C, et al. (2018). "How people domesticated Amazonian forests." *Front. Ecol. Evol.,* https://doi.org/10.3389/fevo.2017.00171; Badiou A (2019). *I Know There Are So Many of You.* The Polity Press, Cambridge.

323. Marx K (1857; 1965). *Pre-Capitalist Economic Formations.* International Publishers, New York; Wood EM (1999; 2002). *The Origins of Capitalism: A Longer View.* Verso, New York; Perelman M (2000). *The Invention of Capitalism.* Duke University Press, Durham, NC; Carlson C (2018). "Rethinking the agrarian question: Agriculture and underdevelopment in the Global South." *Journal of Agrarian Change* 18(4): 703–721.

324. Mészáros I (1970). *Marx's Theory of Alienation*; Arrighi G (1994; 2010). *The Long Twentieth Century: Money, Power and the Origins of Our Times.* Verso, London; Van Bavel BJP (2010). *Manors and Markets: Economy and*

Society in the Low Countries 500-1600. Oxford University Press, Oxford, UK; Ratcliff J (2016). "The great data divergence: Global history of science within global economic history." In P Manning and D Rood (eds), *Global Scientific Practice in an Age of Revolutions, 1750–1850.* University of Pittsburgh Press, Pittsburgh; Rieppel L, W Deringer, and E Lean (2018). *Science and Capitalism: Entangled Histories. Osiris, Volume 33.* University of Chicago Press, Chicago.

325. Rodney W (1972; 2018). *How Europe Underdeveloped Africa.* Verso, New York; Moore, J (2016). "The rise of cheap nature." In J Moore (ed), *Anthropocene or Capitalocene? Nature, History, and the Crisis of Capitalism.* PM Press, Oakland, CA.

326. Ramankutty N, Z Mehrabi, K Waha, L Jarvis, C Kremen, M Herrero, and LH Rieseberg (2018). "Trends in global agricultural land use: Implications for environmental health and food security." *Annu Rev Plant Biol.,* 69: 789–815.

327. Harvey D (1982; 2006). *The Limits to Capital.* Verso, New York; Van Boeckel TP, W Thanapongtharm, T Robinson, L D'Aietti, and M Gilbert (2012). "Predicting the distribution of intensive poultry farming in Thailand." *Agriculture, Ecosystems & Environment,* 149: 144–153; Wallace RG (2016). *Big Farms Make Big Flu: Dispatches on Infectious Disease, Agribusiness, and the Nature of Science.* Monthly Review Press, New York; Bergmann LR and M Holmberg (2016). "Land in motion." *Annals of the American Association of Geographers,* 106(4): 932956; Bergmann LR (2017). "Towards economic geographies beyond the Nature-Society divide." *Geoforum,* 85: 324–335; Holt-Giménez E (2017). *A Foodie's Guide to Capitalism: Understanding the Political Economy of What We Eat.* Monthly Review Press, New York.

328. Lefebvre H (1974; 2000). *The Production of Space.*; Tsing, A (2009). "Supply chains and the human condition." *Rethinking Marxism,* 21(2): 148–176; Patel RAJ and JW Moore (2017). *A History of the World in Seven Cheap Things.* University of California Press, Berkeley; Quintus S and EE Cochrane (2018). "The prevalence and importance of niche construction in agricultural development in Polynesia." *Journal of Anthropological Archaeology,* 51: 173–186; Morrison KD (2018). "Empires as ecosystem engineers: Toward a nonbinary political ecology." *Journal of Anthropological Archaeology,* 52: 196–203; Ficek RE (2019) "Cattle, capital, colonization: Tracking creatures of the Anthropocene in and out of human projects." *Current Anthropology,* 60(S20): S260-S271.

329. Arrighi G (1994; 2010). *The Long Twentieth Century: Money, Power and the Origins of Our Times*; Araghi F (2009). "The invisible hand and the visible foot: Peasants, dispossession and globalization." In AH Akram-Lodhi and C Kay (eds), *Peasants and Globalization: Political Economy, Rural Transformation and the Agrarian Question.* Routledge, New York, pp 111–147; McMichael P (2013). *Food Regimes and Agrarian Questions.* Fernwood Publishing, Black Point, Canada.

330. Arrighi G (1994; 2010). *The Long Twentieth Century: Money, Power and the Origins of Our Times*; Bergmann, LR (2017). "Towards economic geographies beyond the Nature-Society divide"; Zerbe N (2019). "Food as commodity." In JL Vivero-Pol, T Ferrando, O De Schutter, and U Mattei (eds), *Routledge Handbook of Food as a Commons.* Routledge, New York, pp 155–170; Wallace R, LF Chaves, LR Bergmann, C Ayres, L Hogerwerf, R Kock, and RG Wallace (2018). *Clear-Cutting Disease Control: Capital-Led Deforestation, Public Health Austerity, and Vector-Borne Infection.* Springer, Cham.

331. Gilbert M, G Conchedda, TP Van Boeckel, G Cinardi, C Linard, G Nicolas, et al. (2015). "Income disparities and the global distribution of intensively farmed chicken and pigs." *PLoS ONE,* 10(7): e0133381; Fritz S, L See, I McCallum, L You, A Bun, E Moltchanova, M Duerauer, et al. (2015). "Mapping global cropland and field size." *Glob Chang Biol.,* 21(5): 1980–1992.

332. Hoffmann I (2010). "Livestock biodiversity"; Wallace R, L Bergmann, L Hogerwerf, R Kock, and RG Wallace (2016). "Ebola in the hog sector: Modeling pandemic emergence in commodity livestock."

333. Gilbert M, et al. (2015). "Income disparities and the global distribution of intensively farmed chicken and pigs."

334. Goldewijk KK, A Beusen, J Doelman, and E Stehfest (2017). "Anthropogenic land use estimates for the Holocene—HYDE 3.2." *Earth Syst. Sci. Data* 9(2): 927–953.

335. Lassaletta L, F Estellés, AHW.Beusen, L Bouwman, and S Calvet (2019). "Future global pig production systems according to the Shared Socioeconomic Pathways." *Science of the Total Environment,* 665: 739–751.

336. Kreidenweis U, F Humpenöder, L Kehoe, T Kuemmerle, BL Bodirsky, H Lotze-Campen, and A Popp (2018). "Pasture intensification is insufficient to relieve pressure on conservation priority areas in open agricultural markets." *Glob Chang Biol.,* 24(7): 3199–3213.

337. Fritz S., et al. (2015). "Mapping global cropland and field size."

338. White EV and DP Roy (2015). "A contemporary decennial examination of changing agricultural field sizes using Landsat time series data." *Geo.,* 2(1): 33–54.

339. Wallace RG (2018). "Vladimir Iowa Lenin 1: A Bolshevik's study of American agriculture." *Capitalism Nature Socialism* 29(2): 92–107.

340. Wang J, X Xiao, Y Qin, J Dong, G Zhang, W Kou, C Jin, Y Zhou, and Y Zhang (2015). "Mapping paddy rice planting area in wheat-rice double-cropped areas through integration of Landsat-8 OLI, MODIS, and PALSAR images." *Sci Rep.* 5: 10088.

341. Pereira HM, LM Navarro, and IS, Martins (2012). "Global biodiversity change: the bad, the good, and the unknown." *Annu. Rev. Environ. Resourc.* 37: 25–50; Lowder SK, J Skoet, and T Raney (2016). "The number, size, and distribution of farms, smallholder farms, and family farms worldwide." *World Dev.* 87: 16–29.

342. Ramankutty N, et al. (2018). "Trends in global agricultural land use: Implications for environmental health and food security."

343. Haesbaert R (2011). *El Mito de la Desterritorializacíon: Del Fin de los Territorios a la Multiterritorialidad* [The Myth of De-territorialization: From the End of Territories to Multiterritoriality]. México: Siglo XXI. www. scielo.org.mx/pdf/crs/v8n15/v8n15a1.pdf; Craviotti C (2016). "Which territorial embeddedness? Territorial relationships of recently internationalized firms of the soybean chain." *Journal of Peasant Studies*, 43(2): 331–347; Wallace R, et al. (2018). *Clear-Cutting Disease Control: Capital-Led Deforestation, Public Health Austerity, and Vector-Borne Infection.*

344. Jepson W, C Brannstrom, and A Filippi (2010). "Access regimes and regional land change in the Brazilian Cerrado, 1972–2002." *Annals of the Association of American Geographers* 100(1): 87–111; Garrett RD, EF Lambin, and RL Naylor (2013). "The new economic geography of land use change: Supply chain configurations and land use in the Brazilian Amazon." *Land Use Policy* 34: 265–275; Meyfroidt P, KM Carlson, MF Fagan, VH Gutiérrez-Vlez, MN Macedo, et al. (2014). "Multiple pathways of commodity crop expansion in tropical forest landscapes." *Environ. Res. Lett.* 9:074012; Geldes C, C Felzensztein, E Turkina, and A Durand (2015). "How does proximity affect interfirm marketing cooperation? A study of an agribusiness cluster." *Journal of Business Research* 68(2): 263–272; Oliveira GLT (2016). "The geopolitics of Brazilian soybeans." *The Journal of Peasant Studies* 43(2): 348–372; Godar J, C Suavet, TA Gardner, E Dawkins, and P Meyfroidt (2016). "Balancing detail and scale in assessing transparency to improve the governance of agricultural commodity supply chains." *Environ. Res. Lett.* 11: 035015.

345. Turzi M (2011). "The soybean republic." *Yale Journal of International Affairs* 6(2): 5968; Harvey D (2018). "Marx's refusal of the labour theory of value: Reading Marx's *Capital* with David Harvey." 14 March. http://davidharvey.org/2018/03/marxsrefusal-of-the-labour-theory-of -value-by-david-harvey/.

346. Araghi F (2009). "The invisible hand and the visible foot: Peasants, dispossession and globalization"; Mastrangelo ME and S Aguiar (2019). "Are ecological modernization narratives useful for understanding and steering social-ecological change in the Argentine Chaco?" *Sustainability* 11(13): 3593; Wallace R, A Liebman, L Bergmann, and RG Wallace (2020). "Agribusiness vs. public health: Disease control in resource-asymmetric conflict." https://hal.archives-ouvertes.fr/hal-02513883/.

347. Wallace RG (2009). "Breeding influenza: the political virology of offshore farming." *Antipode* 41: 916–951; Messinger S and A Ostling (2009). "The consequences of spatial structure for pathogen evolution." *The American Naturalist* 174: 441–454; Atkins KE, RG Wallace, L Hogerwerf, M Gilbert, J Slingenbergh, J Otte, and A Galvani (2010). *Livestock Landscapes and the Evolution of Influenza Virulence. Virulence Team Working Paper*

No. 1. Animal Health and Production Division, Food and Agriculture Organization of the United Nations, Rome; Atkins KE, AF Read, NJ Savill, KG Renz, AF Islam, SW Walkden-Brown, and ME Woolhouse (2013). "Vaccination and reduced cohort duration can drive virulence evolution: Marek's disease virus and industrialized agriculture." *Evolution* 67(3): 851–860; Wallace R and RG Wallace (2015). "Blowback: new formal perspectives on agriculturally-driven pathogen evolution and spread." *Epidemiology and Infection* 143(10): 2068–2080.

348. Wallace R, et al. (2018). *Clear-Cutting Disease Control: Capital-Led Deforestation, Public Health Austerity, and Vector-Borne Infection*; Porter N (2019). *Viral Economies: Bird Flu Experiments in Vietnam.* The University of Chicago Press, Chicago; Wallace RG, R Alders, R Kock, T Jonas, R Wallace, and L Hogerwerf (2019). "Health before medicine: Community resilience in food landscapes." In M Walton (ed), *One Planet, One Health: Looking After Humans, Animals and the Environment.* Sydney University Press, Sydney.

349. Gunderson R (2011). "The metabolic rifts of livestock agribusiness." *Organization & Environment,* 24(4): 404–422; Gunderson R (2013). "From cattle to capital: Exchange value, animal commodification, and barbarism." *Critical Sociology,* 39(2): 259–275.

350. Novek, J (2003). "Intensive livestock operations, disembedding, and polarization in Manitoba." *Society & Natural Resources,* 16(7): 567–581; Clausen R and B Clark (2005). "The metabolic rift and marine ecology: An analysis of the ocean crisis within capitalist production." *Organization & Environment,* 18(4): 422–444; Mancus P (2007). "Nitrogen fertilizer dependency and its contradictions: A theoretical exploration of social-ecological metabolism." *Rural Sociology,* 72(2): 269–288; Foster JB, R York, and B Clark (2010). *The Ecological Rift: Capitalism's War on the Earth.* Monthly Review Press, New York; Schneider M (2017). "Wasting the rural: Meat, manure, and the politics of agro-industrialization in contemporary China." *Geoforum,* 78: 89–97.

351. Foster JB (2016). "Marx as a food theorist." *Monthly Review,* 68(7): 1–22.

352. Lewontin R (1998). "The maturing of capitalist agriculture: Farmer as proletarian." *Monthly Review,* 50(3): 72; Moore J (2003). "Capitalism as World-Ecology: Braudel and Marx on environmental history." *Organization & Environment,* 16: 431–458; Akram-Lodhi AH and C Kay (eds) (2009). *Peasants and Globalization: Political Economy, Rural Transformation and the Agrarian Question.* Routledge, New York; Bernstein H (2010). *Class Dynamics of Agrarian Change.* A. Kumarian Press; Gunderson R (2011). "Marx's comments on animal welfare." *Rethinking Marxism,* 23(4): 543–548; Banaji J (2016). "Merchant capitalism, peasant households and industrial accumulation: Integration of a model." *Journal of Agrarian Change,* 16(3): 410–431; Okamoto, KW, A Liebman, and RG Wallace (2020) "At what geographic scales does agricultural alienation amplify foodborne disease outbreaks? A statistical test for 25 U.S. States, 1970–2000."

353. Marsden T (2016). "Exploring the rural eco-economy: Beyond neoliberalism." *Sociologia Ruralis*, 56(4): 597–615; Weber MB (2018). *Manufacturing the American Way of Farming: Agriculture, Agribusiness, and Marketing in the Postwar Period.* Dissertation. Department of History, Iowa State University. https://lib.dr.iastate.edu/etd/16485/; Peng B, Z Liu, B Zhang, and X Chen (2018). "Idyll or nightmare: what does rurality mean for farmers in a Chinese village undergoing commercialization?" *Inter-Asia Cultural Studies*, 19(2): 234–251; Tonts M and J Horsley (2019). "The neoliberal countryside." In M Scott, N Gallent, and M Gkartzios (eds), *The Routledge Companion to Rural Planning*. Routledge, New York.

354. Patton D (2018). "China's multi-story hog hotels elevate industrial farms to new levels." Reuters, 10 May. https://www.reuters.com/article/us-china-pigs-hotels-insight/insight-chinas-multi-story-hog-hotels-elevate-industrial-farms-to-new-levels-idUSKBN1IB362.

355. Arrighi G (1994; 2010). *The Long Twentieth Century: Money, Power and the Origins of Our Times*; Ministry of Economic Affairs (2017). "Agri & food exports achieve record high in 2016." Government of the Netherlands, news release, January 20. https://www.government.nl/latest/news/2017/01/20/agri-foodexports-achieve-record-high-in-2016; Wallace RG, A Liebman, and L Bergmann (2018). *Are Dutch Livestock Rifts Founded upon Relational Geographies Past and Present?* Report for the Centre for Infectious Disease Control of the National Institute for Public Health and the Environment, the Netherlands.

356. Berry W (2015). "Farmland without farmers." *The Atlantic*. 19 March. https://www.theatlantic.com/national/archive/2015/03/farmland-without-farmers/388282/; Cromartie J (2017). *Rural Areas Show Overall Population Decline and Shifting Regional Patterns of Population Change*. USDA Economic Research Service. 5 September. https://ageconsearch.umn.edu/record/265963/files/https_www_ers_usda_gov_amber-waves_2017_september_rural-areas-show-overall-population-decline-and-shifting-regional-patterns-of-population-c.pdf.

357. Jones CS, CW Drake, CE Hruby, KE Schilling, and CF Wolter (2019). "Livestock manure driving stream nitrate." *Ambio* 48(10): 1143–1153.

358. Jones CS (2019). "Iowa's real population." *Chris Jones, IIHR Research Engineer* blog. https://www.iihr.uiowa.edu/cjones/iowas-real-population/.

359. Merchant JA, AL Naleway, ER Svendsen, KM Kelly, LF Burmeister, et al. (2005). "Asthma and farm exposures in a cohort of Rural Iowa children." *Environ Health Perspect* 113: 350–356; Kleinschmidt TL (2011). *Modeling Hydrogen Sulfide Emissions: Are Current Swine Animal Feeding Operation Regulations Effective at Protecting against Hydrogen Sulfide Exposure in Iowa?* Master of Science thesis. Department of Occupational and Environmental Health, University of Iowa. https://ir.uiowa.edu/cgi/viewcontent.cgi?article=2708&context=etd; Pavilonis BT, WT Sanderson, and JA Merchant (2013). "Relative exposure to swine animal feeding operations

and childhood asthma prevalence in an agricultural cohort." *Environmental Research* 122: 74–80; Schechinger A (2019). *Contamination of Iowa's Private Wells: Methods and Detailed Results.* Environmental Working Group and Iowa Environmental Council. https://www.ewg.org/iowawellsmethods; Tessum CW, JS Apte, AL Goodkind, NZ Muller, KA Mullins, et al. (2019). "Inequity in consumption of goods and services adds to racial–ethnic disparities in air pollution exposure." *PNAS* 116(13): 6001–6006.

360. Roe B, EG Irwin, and JS Sharp (2002). "Pigs in space: Modeling the spatial structure of hog production in traditional and nontraditional production regions." *Amer. J. Agr. Econ.* 84(2): 259–278; Du S (2018). "These rookie Minnesota farmers fight with Big Ag." *City Pages*, 22 August. http://www.citypages.com/news/these-rookie-minnesota-farmers-are-picking-a-fight-with-big-ag/491393901; Wilcox J (2019). "Hog confinement no longer coming to Fillmore County." KIMT News, 13 February. https://www.kimt.com/content/news/Hog-Confinement-no-longer-coming-to-Fillmore-County-505810161.html; Kaeding D (2019). "Proposed hog farm stirs up debate in northern Wisconsin." *Duluth News-Tribune*, 23 April. https://www.duluthnewstribune.com/business/4602936-proposed-hog-farm-stirs-debate-northern-wisconsin.

361. Friedmann H (2005). "From colonialism to green capitalism: Social movements and emergence of food regimes." In F Buttel and P McMichael (eds), *New Directions in the Sociology of Global Development (Research in Rural Sociology and Development, Vol. 11).* Emerald Group Publishing Limited, Bingley, pp 227–264; McMichael P (2009). "A food regime genealogy." *Journal of Peasant Studies* 36(1): 139–169; Wilkie RM (2012). *Livestock/ Deadstock: Working with Farm Animals from Birth to Slaughter.* Temple University Press, Philadelphia; Blanchette A (2018). "Industrial meat production." *Annual Review of Anthropology* 47: 185–199.

362. Watts MJ (2004). "Are hogs like chickens? Enclosure and mechanization in two 'white meat' filières." In A Hughes and S Reimer (eds), *Geographies of Commodity Chains.* Routledge, London, pp 39–62; Shukin N (2009). *Animal Capital: Rendering Life in Biopolitical Times.* University of Minnesota Press, Minneapolis; Hendrickson MK (2015). "Resilience in a concentrated and consolidated food system." *Journal of Environmental Studies and Sciences* 5(3): 418–431; Beldo L (2017). "Metabolic labor: Broiler chickens and the exploitation of vitality." *Environmental Humanities* 9(1): 108–128; Gisolfi MR (2017). *The Takeover: Chicken Farming and the Roots of American Agribusiness.* The University of Georgia Press, Athens, GA.

363. Le Rouzic A, JM Álvarez-Castro, and Ö Carlborg (2008). "Dissection of the genetic architecture of body weight in chicken reveals the impact of epistasis on domestication traits." *Genetics*, 179(3): 1591–1599; Hill WG and M Kirkpatrick (2010). "What animal breeding has taught us about evolution." *Annual Review of Ecology, Evolution, and Systematics* 41: 1–19; Komiyama T, M Lin, and A Ogura (2016). "aCGH analysis to estimate genetic

variations among domesticated chickens." *BioMed Research International* 2016:1794329; Wallace RG (2016). *Big Farms Make Big Flu: Dispatches on Infectious Disease, Agribusiness, and the Nature of Science;* Isik F, J Holland, and C Maltecca (2017). *Genetic Data Analysis for Plant and Animal Breeding.* Springer, Cham.

364. Watts MJ (2004). "Are hogs like chickens? Enclosure and mechanization in two 'white meat' filières"; Genoways T (2014). *The Chain: Farm, Factory, and the Fate of our Food.* Harper, New York; Leonard C (2014). *The Meat Racket: The Secret Takeover of America's Food Business.* Simon & Schuster, New York; Wallace RG (2016). *Big Farms Make Big Flu: Dispatches on Infectious Disease, Agribusiness, and the Nature of Science;* Blanchette A (2018). "Industrial meat production."

365. Johnson NL (1995). *The Diffusion of Livestock Breeding Technology in the U.S.: Observations on the Relationship between Technical Change and Industry Structure.* Staff Paper Series P95-11. Department of Applied Economics. College of Agricultural, Food, and Environmental Sciences, University of Minnesota; Schmidt TP (2017). *The Political Economy of Food and Finance.* Routledge, New York; Bjorkhaug H, A Magnan, and G Lawrence (eds) (2018). *The Financialization of Agri-Food Systems: Contested Transformations.* Routledge, New York.

366. Fuglie KO, PW Heisey, JL King, K Day-Rubenstein, D Schimmelpfennig, and SL Wang (2011). *Research Investments and Market Structure in Food Processing, Agricultural Input, and Biofuel Industries Worldwide.* Economic Information Bulletin Number 90. Economic Research Service. USDA; MacDonald JM, R Hoppe, and D Newton (2018). *Three Decades of Consolidation in U.S. Agriculture.* Economic Information Bulletin Number 189. Economic Research Service, USDA; Neo H and J Emel (2017). *Geographies of Meat: Politics, Economy and Culture.* Routledge, New York; Schmidt TP (2017). *The Political Economy of Food and Finance.*

367. Gura S (2007). *Livestock Genetics Companies. Concentration and Proprietary Strategies of an Emerging Power in the Global Food Economy.* League for Pastoral Peoples and Endogenous Livestock Development, Ober-Ramstadt, Germany. http://www.pastoralpeoples.org/docs/livestock_genetics_en.pdf.

368. Johnson NL (1995). *The Diffusion of Livestock Breeding Technology in the U.S.: Observations on the Relationship between Technical Change and Industry Structure.*

369. Neo H and J Emel (2017). *Geographies of Meat: Politics, Economy and Culture.*

370. Bugos GE (1992). "Intellectual property protection in the American chicken-breeding industry." *Business History Review* 66: 127–168; Narrod CA and KO Fuglie (2000). "Private investment in livestock breeding with implications for public research policy." *Agribusiness* 16(4): 457–470; Koehler-Rollefson I (2006). "Concentration in the poultry sector." Presentation at "The Future of Animal Genetic Resources: Under Corporate

Control or in the Hands of Farmers and Pastoralists?" International workshop, Bonn, Germany, 16 October. http://www.pastoralpeoples.org/docs/03Koehler-RollefsonLPP.pdf.

371. MacDonald JM, R Hoppe, and D Newton (2018). *Three Decades of Consolidation in U.S. Agriculture*; Wainwright, W (2018). *Economic Instruments for Supplying Agrobiodiversity Conservation.* Dissertation, Department of GeoSciences, University of Edinburgh.

372. Arthur JA and GAA Albers (2003). "Industrial perspective on problems and issues associated with poultry breeding." In WM Muir and SE Aggrey (eds), *Poultry Genetics, Breeding and Biotechnology.* CABI Publishing, UK; Whyte J, E Blesbois, and MJ McGrew (2016). "Increased sustainability in poultry production: New tools and resources for genetic management." In E Burton, et al. (eds), *Sustainable Poultry Production in Europe.* Poultry Science Symposium Series, Vol. 31. CABI, Wallingford, UK.

373. Fulton JE (2006). "Avian genetic stock preservation: An industry perspective." *Poultry Science,* 85(2): 227–231.

374. Blackburn HD (2006). "The National Animal Germplasm Program: Challenges and opportunities for poultry genetic resources." *Poultry Science,* 85(2): 210–215; Hoffmann I (2009). "The global plan of action for animal genetic resources and the conservation of poultry genetic resources." *World's Poultry Science Journal* 65(2): 286–297; Hoffmann I (2010). "Livestock biodiversity"; Wainwright W (2018). *Economic Instruments for Supplying Agrobiodiversity Conservation.*

375. Rubin CJ, et al. (2010). "Whole-genome resequencing reveals loci under selection during chicken domestication." *Nature* 464(7288): 587–91.

376. Zuidhof MJ, BL Schneider, VL Carney, DR Korver, and FE Robinson (2014). "Growth, efficiency, and yield of commercial broilers from 1957, 1978, and 2005." *Poultry Science* 93(12): 2970–2982.

377. Knowles TG, SC Kestin, SM Haslam, SN Brown, LE Green, et al. (2008). "Leg disorders in broiler chickens: Prevalence, risk factors and prevention." *PLoS ONE* 3(2): e1545.

378. González LA, X Manteca, S Calsamiglia, KS Schwartzkopf-Genswein, and A Ferret (2012). "Ruminal acidosis in feedlot cattle: Interplay between feed ingredients, rumen function and feeding behavior (a review)." *Animal Feed Science and Technology* 172(1-2): 66–79.

379. Okamoto K, A Liebman, and RG Wallace (2019). "At what geographic scales does agricultural alienation amplify foodborne disease outbreaks? A statistical test for 25 U.S. states, 1970–2000"; Wallace RG, K Okamoto, and A Liebman (2020). "Gated ecologies."

380. Casalone C and J Hope (2018). "Atypical and classic bovine spongiform encephalopathy." *Handbook of Clinical Neurology* 153: 121–134.

381. Simmons M, G Ru, C Casalone, B Iulini, C Cassar, and T Seuberlich (2018). "DISCONTOOLS: Identifying gaps in controlling bovine spongiform encephalopathy." *Transboundary and Emerging Diseases* 65(S1): 9–21.

382. Travel A, Y Nys, and M Bain (2011). "Effect of hen age, moult, laying environment and egg storage on egg quality." In Y Nys, M Bain, and F Van Immerseel (eds), *Improving the Safety and Quality of Eggs and Egg Products: Egg Chemistry, Production and Consumption*. Woodhead Publishing, pp 300–329; Wolc A, J Arango, T Jankowski, I Dunn, P Settar, et al. (2014). "Genome-wide association study for egg production and quality in layer chickens." *Journal of Animal Breeding and Genetics* 131(3): 173–182; Bédécarrats GY and C Hanlon (2017). "Effect of lighting and photoperiod on chicken egg production and quality." In PY Hester (ed), *Egg Innovations and Strategies for Improvements*. Academic Press, pp 65–75.

383. Ross KA, AD Beaulieu, J Merrill, G Vessie, and JF Patience (2011). "The impact of ractopamine hydrochloride on growth and metabolism, with special consideration of its role on nitrogen balance and water utilization in pork production." *J Anim Sci.*, 89(7): 2243–2256; Liu X, DK Grandy, and A Janowsky (2014). "Ractopamine, a livestock feed additive, is a full agonist at trace amine-associated receptor 1." *J Pharmacol Exp Ther.* 350(1): 124–129; Strom S (2015) "New type of drug-free labels for meat Has U.S.D.A. blessing." *New York Times*, 4 September. https://www.nytimes.com/2015/09/05/business/new-type-of-drug-free-labels-for-meat-has-usda-blessing.html; Wallace RG (2015). "Eating the brown acid." *Farming Pathogens*. 21 September. https://farmingpathogens.wordpress.com/2015/09/21/eating-the-brown-acid/.

384. Bottelmiller H (2012). "Dispute over drug in feed limits U.S. meat exports." Food & Environment Reporting Network. 25 January. https://thefern.org/2012/01/dispute-over-drug-in-feed-limiting-u-s-meat-exports/.

385. Graham JP, JH Leibler, LB Price, JM Otte, DU Pfeiffer, T Tiensin, and EK Silbergeld. (2008). "The animal–human interface and infectious disease in industrial food animal production: Rethinking biosecurity and biocontainment." *Public Health Reports* 123: 28–299; Hincliffe S (2013). "The insecurity of biosecurity: remaking emerging infectious diseases." In A Dobson, K Baker, and SL Taylor (eds), *Biosecurity: The Socio-Politics of Invasive Species and Infectious Diseases*. Routledge, New York, pp 199–214; Allen J and S Lavau (2015). "'Just-in-time' disease: Biosecurity, poultry and power." *Journal of Cultural Economy* 8(3): 342–360; Wallace RG (2016). *Big Farms Make Big Flu*; Leibler JH, K Dalton, A Pekosz, GC Gray, and EK Silbergeld (2017). "Epizootics in industrial livestock production: Preventable gaps in biosecurity and biocontainment." *Zoonoses Public Health* 64(2): 137–145.

386. Tauxe RV (1997). "Emerging foodborne diseases: An evolving public health challenge." *Emerging Infectious Diseases* 3(4): 425–434; Guinat C, A Gogin, S Blome, G Keil, R Pollin, et al. (2016). "Transmission routes of African swine fever virus to domestic pigs: current knowledge and future research directions." *Vet Rec.* 178(11): 262–267; Wallace R, L Bergmann, L Hogerwerf, R Kock, and RG Wallace (2016). "Ebola in the hog sector: Modeling pandemic emergence in commodity livestock"; Marder EP, PM Griffin, PR Cieslak, J Dunn, S Hurd, et al. (2018). "Preliminary incidence

and trends of infections with pathogens transmitted commonly through food—Foodborne Diseases Active Surveillance Network, 10 U.S. sites, 2006–2017." *MMWR*, 67(11): 324–328; Wallace RG, K Okomoto, and A Liebman (2020). "Gated ecologies"; Wallace RG (2020). "Notes on a novel coronavirus." This volume.

387. Garrett KA and CM Cox (2008). "Applied biodiversity science: Managing emerging diseases in agriculture and linked natural systems using ecological principles." In RS Ostfeld, F Keesing and VT Eviner (eds), *Infectious Disease Ecology: Effects of Ecosystems on Disease and of Disease on Ecosystems.* Princeton University Press, Princeton, pp 368–386; Vandermeer J (2010). *The Ecology of Agroecosystems.* Jones and Bartlett Publishers, Sudbury, MA; Thrall PH, JG Oakeshott, G Fitt, S Southerton, JJ Burdon, et al. (2011). "Evolution in agriculture—the application of evolutionary approaches to the management of biotic interactions in agro-ecosystems." *Evolutionary Applications* 4: 200–215; Denison RF (2012). *Darwinian Agriculture: How Understanding Evolution Can Improve Agriculture.* Princeton University Press, Princeton, NJ; Gilbert M, X Xiao, and TP Robinson (2017). "Intensifying poultry production systems and the emergence of avian influenza in China: A 'One Health/Ecohealth' epitome." *Archives of Public Health* 75.

388. Houshmand M, K Azhar, I Zulkifli, MH Bejo, and A Kamyab (2012). "Effects of prebiotic, protein level, and stocking density on performance, immunity, and stress indicators of broilers." *Poult Sci.*, 91: 393–401; Gomes AVS, WM Quinteiro-Filho, A Ribeiro, V Ferraz-de-Paula, ML Pinheiro, et al. (2014). "Overcrowding stress decreases macrophage activity and increases Salmonella Enteritidis invasion in broiler chickens." *Avian Pathology* 43(1): 82–90; Yarahmadi P, HK Miandare, S Fayaz, C Marlowe, and A Caipang (2016). "Increased stocking density causes changes in expression of selected stress- and immune-related genes, humoral innate immune parameters and stress responses of rainbow trout (*Oncorhynchus mykiss*)." *Fish & Shellfish Immunology* 48: 43–53; Li W, F Wei, B Xu, Q Sun, W Deng, et al. (2019). "Effect of stocking density and alpha-lipoic acid on the growth performance, physiological and oxidative stress and immune response of broilers." *Asian-Australasian Journal of Animal Studies.* https://doi.org/10.5713/ajas.18.0939.

389. Pitzer VE, R Aguas, S Riley, WLA Loeffen, JLN Wood, and BT Grenfell (2016). "High turnover drives prolonged persistence of influenza in managed pig herds." *J. R. Soc. Interface* 13: 20160138; Gast RK, R Guraya, DR Jones, KE Anderson, and DM Karcher (2017). "Frequency and duration of fecal shedding of *Salmonella* Enteritidis by experimentally infected laying hens housed in enriched colony cages at different stocking densities." *Front. Vet. Sci.* https://doi.org/10.3389/fvets.2017.00047; Diaz A, D Marthaler, C Corzo, C Muñoz-Zanzi, S Sreevatsan, M Culhane, and M Torremorell (2017). "Multiple genome constellations of similar and distinct influenza

A viruses co-circulate in pigs during epidemic events." *Scientific Reports* 7: 11886; EFSA Panel on Biological Hazards (EFSA BIOHAZ Panel), K Koutsoumanis, A Allende, A Alvarez-Ordóñez, D Bolton, et al. (2019). "*Salmonella* control in poultry flocks and its public health impact." *EFSA Journal* 17(2): e05596.

390. Atkins KE, RG Wallace, L Hogerwerf, M Gilbert, J Slingenbergh, J Otte, and A Galvani (2011). *Livestock Landscapes and the Evolution of Influenza Virulence. Virulence Team Working Paper No. 1.* Animal Health and Production Division, Food and Agriculture Organization of the United Nations, Rome; Allen J and S Lavau (2015). "'Just-in-time' disease: Biosecurity, poultry and power"; Pitzer VE, R Aguas, S Riley, WLA Loeffen, JLN Wood, and BT Grenfell (2016). "High turnover drives prolonged persistence of influenza in managed pig herds"; Rogalski MA, CD Gowler, CL Shaw, RA Hufbauer, and MA Duffy (2017). "Human drivers of ecological and evolutionary dynamics in emerging and disappearing infectious disease systems." *Phil. Trans. R. Soc. B* 372(1712): 20160043.

391. Rogalski MA, CD Gowler, CL Shaw, RA Hufbauer, and MA Duffy (2017). "Human drivers of ecological and evolutionary dynamics in emerging and disappearing infectious disease systems."

392. Rozins C and T Day (2017). "The industrialization of farming may be driving virulence evolution." *Evolutionary Applications* 10(2): 189–198.

393. Wallace RG (2009). "Breeding influenza: the political virology of offshore farming"; Atkins KE, AF Read, NJ Savill, KG Renz, AF Islam, SW Walkden-Brown, and ME Woolhouse (2013). "Vaccination and reduced cohort duration can drive virulence evolution: Marek's disease virus and industrialized agriculture." *Evolution* 67(3): 851–860; Wallace RG (2016). "Flu the farmer." In *Big Farms Make Big Flu: Dispatches on Infectious Disease, Agribusiness, and the Nature of Science.* Monthly Review Press, New York, pp 316–318; Mennerat A, MS Ugelvik, CH Jensen, and A Skorping (2017). "Invest more and die faster: The life history of a parasite on intensive farms." *Evolutionary Applications* 10(9): 890–896.

394. Atkins KE, RG Wallace, L Hogerwerf, M Gilbert, J Slingenbergh, J Otte, and A Galvani (2011). *Livestock Landscapes and the Evolution of Influenza Virulence;* Kennedy DA, C Cairns, MJ Jones, AS Bell, RM Salathe, et al. (2017). "Industry-wide surveillance of Marek's disease virus on commercial poultry farms." *Avian Dis.* 61: 153–164.

395. Wallace RG (2016). "A pale, mushy wing." In *Big Farms Make Big Flu: Dispatches on Infectious Disease, Agribusiness, and the Nature of Science.* Monthly Review Press, New York, pp 222–223; Gilbert M, X Xiao, and TP Robinson (2017). "Intensifying poultry production systems and the emergence of avian influenza in China: A 'One Health/Ecohealth' epitome."

396. Wallace RG (2009). "Breeding influenza: the political virology of offshore farming"; Atkins KE, RG Wallace, L Hogerwerf, M Gilbert, J Slingenbergh, J Otte, and A Galvani (2011). *Livestock Landscapes and the Evolution of*

Influenza Virulence; Dhingra MS, J Artois, S Dellicour, P Lemey, G Dauphin, et al. (2018). "Geographical and historical patterns in the emergences of novel Highly Pathogenic Avian Influenza (HPAI) H5 and H7 viruses in poultry." *Front. Vet. Sci.* 05. https://doi.org/10.3389/fvets.2018.00084.

397. Nelson MI, P Lemey, Y Tan, A Vincent, TT Lam, et al. (2011). "Spatial dynamics of human-origin H1 influenza A virus in North American swine." *PLoS Pathog.* 7(6):e1002077; Fuller TL, M Gilbert, V Martin, J Cappelle, P Hosseini, KY Njabo, S Abdel Aziz, X Xiao, P Daszak, and TB Smith (2013). "Predicting hotspots for influenza virus reassortment." *Emerg Infect Dis.* 19(4): 581–588; Wallace R and RG Wallace (2015). "Blowback: new formal perspectives on agriculturally-driven pathogen evolution and spread." *Epidemiology and Infection* 143(10): 2068–2080; Mena I, MI Nelson, F Quezada-Monroy, J Dutta, R Cortes-Fernández, JH Lara-Puente, F Castro-Peralta, LF Cunha, NS Trovão, B Lozano-Dubernard, A Rambaut, H van Bakel, and A García-Sastre (2016). "Origins of the 2009 H1N1 influenza pandemic in swine in Mexico." *Elife* 5.pii:e16777; O'Dea EB, H Snelson, and S Bansal (2016). "Using heterogeneity in the population structure of U.S. swine farms to compare transmission models for porcine epidemic diarrhea." *Scientific Reports* 6: 22248; Dee SA, FV Bauermann, MC Niederwerder, A Singrey, T Clement, et al. (2018). "Survival of viral pathogens in animal feed ingredients under transboundary shipping models." *PLoS ONE* 14(3): e0214529; Gorsich EE, RS Miller, HM Mask, C Hallman, K Portacci, and CT Webb (2019). "Spatio-temporal patterns and characteristics of swine shipments in the U.S. based on Interstate Certificates of Veterinary Inspection." *Scientific Reports*, 9: 3915; Nelson MI, CK Souza, NS Trovão, A Diaz, I Mena, et al. (2019). "Human-origin influenza A(H3N2) reassortant viruses in swine, Southeast Mexico." *Emerg Infect Dis.* 25(4): 691–700.

398. Rabsch W, BM Hargis, RM Tsolis, RA Kingsley, KH Hinz, H Tschäpe, and AJ Bäumler (2000). "Competitive exclusion of *Salmonella enteritidis* by *Salmonella gallinarum* in poultry." *Emerg Infect Dis.* 6(5): 443–448; Shim E and AP Galvani (2009). "Evolutionary repercussions of avian culling on host resistance and influenza virulence." *PLoS ONE* 4(5): e5503; Nfon C, Y Berhane, J Pasick, C Embury-Hyatt, G Kobinger, et al. (2012). "Prior infection of chickens with H1N1 or H1N2 Avian Influenza elicits partial heterologous protection against Highly Pathogenic H5N1." *PLoS ONE* 7(12): e51933; Yang Y, G Tellez, JD Latorre, PM Ray, X Hernandez, et al. (2018). "Salmonella excludes salmonella in poultry: Confirming an old paradigm using conventional and barcode-tagging approaches." *Front. Vet. Sci.* 5: 101.

399. Smith GJ, XH Fan, J Wang, KS Li, K Qin, et al. (2006). "Emergence and predominance of an H5N1 influenza variant in China." *Proc Natl Acad Sci U S A.* 103(45): 16936–16941; Pasquato A and NG Seidah (2008). "The H5N1 influenza variant Fujian-like hemagglutinin selected following vaccination

exhibits a compromised furin cleavage: neurological consequences of highly pathogenic Fujian H5N1 strains." *J Mol Neurosci.* 35(3): 339–343; Lauer D, S Mason, B Akey, L Badcoe, D Baldwin, et al. (2015). *Report of the Committee on Transmissible Diseases of Poultry and Other Avian Species.* United States Animal Health Association. https://www.usaha.org/upload/ Committee/TransDisPoultry/report-pad-2015.pdf.

400. Wallace RG (2016). "Made in Minnesota." In *Big Farms Make Big Flu: Dispatches on Infectious Disease, Agribusiness, and the Nature of Science.* Monthly Review Press, New York, pp 347–358; Lantos PM, K Hoffman, M Höhle, B Anderson, and GC Gray (2016). "Are people living near modern swine production facilities at increased risk of influenza virus infection?" *Clinical Infectious Diseases* 63(12): 1558–1563; Ma J, H Shen, C McDowell, Q Liu, M Duff, et al. (2019). "Virus survival and fitness when multiple genotypes and subtypes of influenza A viruses exist and circulate in swine." *Virology* 532: 30–38.

401. Kennedy DA, PA Dunn, and AF Read (2018). "Modeling Marek's disease virus transmission: A framework for evaluating the impact of farming practices and evolution." *Epidemics* 23: 85–95.

402. Rozins C and T Day (2017). "The industrialization of farming may be driving virulence evolution." *Evolutionary Applications* 10(2): 189–198; Rozins C, T Day, and S Greenhalgh (2019). "Managing Marek's disease in the egg industry." *Epidemics* 27: 52–58.

403. Bryant L and B Garnham (2014). "Economies, ethics and emotions: Farmer distress within the moral economy of agribusiness." *Journal of Rural Studies* 34: 304–312; Wallace RG (2016). "Collateralized farmers." In *Big Farms Make Big Flu: Dispatches on Infectious Disease, Agribusiness, and the Nature of Science.* Monthly Review Press, New York, pp 336–340; Wallace RG (2017). "Industrial production of poultry gives rise to deadly strains of bird flu H5Nx." Institute for Agriculture and Trade Policy blog, January 24. https://www.iatp.org/blog/201703/industrial-productionpoultry-gives-rise-deadly-strains-bird-flu-h5nx; Wallace RG (2018). *Duck and Cover: Epidemiological and Economic Implications of Ill-founded Assertions that Pasture Poultry Are an Inherent Disease Risk.* Australian Food Sovereignty Alliance. https://afsa.org.au /wpcontent/uploads/2 018/10/WallaceDuck-and-CoverReport-September2018.pdf .

404. Forster P and O Charnoz (2013). "Producing knowledge in times of health crises: Insights from the international response to avian influenza in Indonesia." *Revue d'anthropologie des connaissances* 7(1):w-az; Wallace RG (2016). "A pale, mushy wing."

405. Ingram A (2013). "Viral geopolitics: biosecurity and global health governance." In A Dobson, K Baker, and SL Taylor (eds), *Biosecurity: The Socio-Politics of Invasive Species and Infectious Diseases.* Routledge, New York, pp 137–150; Dixon MW (2015) "Biosecurity and the multiplication of crises in the Egyptian agri-food industry." *Geoforum* 61:90–100.

406. Wallace RG (2018). *Duck and Cover: Epidemiological and Economic Implications of Ill-founded Assertions that Pasture Poultry Are an Inherent Disease Risk.*

407. Wallace RG (2009). "Breeding influenza: the political virology of offshore farming"; Atkins KE, RG Wallace, L Hogerwerf, M Gilbert, J Slingenbergh, J Otte, and A Galvani (2010). *Livestock Landscapes and the Evolution of Influenza Virulence;* Leonard C (2014). *The Meat Racket: The Secret Takeover of America's Food Business.*

408. Lulka D (2004). "Stabilizing the herd: Fixing the identity of nonhumans." *Environment and Planning D* 22(3): 439–463; Lorimer J and C Driessen (2013). "Bovine biopolitics and the promise of monsters in the rewilding of Heck cattle." *Geoforum* 48: 249–259.

409. Harris DL (2000; 2008) *Multi-Site Pig Production.* John Wiley & Sons, Hoboken, NJ.

410. Henry DP (1965). "Experiences during the first eight weeks of life of HYPAR piglets." *Australian Veterinary Journal* 41(5); Harris DL (2000; 2008). *Multi-Site Pig Production;* Stibbe A (2012). *Animals Erased: Discourse, Ecology, and Reconnection with the Natural World.* Wesleyan University Press, Middletown, CT.

411. Muñoz A, G Ramis, FJ Pallarés, JS Martínez, J Oliva, et al. (1999). "Surgical procedure for Specific Pathogen Free piglet by modified terminal hysterectomy." *Transplantation Proceedings* 31: 2627–2629; Zimmerman JJ, LA Karriker, A Ramirez, KJ Schwartz, GW Stevenson, and J Zhang (eds) (2019). *Diseases of Swine.* John Wiley & Sons, Hoboken, NJ.

412. Cameron RDA (2000). *A Review of the Industrialisation of Pig Production Worldwide with Particular Reference to the Asian Region.* http://www.fao. org/ag/againfo/themes/documents/pigs/A%20review%20of%20the%20 industrialisation%20of%20pig%20production%20worldwide%20with%20 particular%20reference%20to%20the%20Asian%20region.pdf; Huang Y, DM Haines, and JCS Harding (2013). "Snatch-farrowed, porcine-colostrum-deprived (SF-pCD) pigs as a model for swine infectious disease research." *Can J Vet Res.,* 77(2): 81–88.

413. Zimmerman JJ, LA Karriker, A Ramirez, KJ Schwartz, GW Stevenson, and J Zhang (eds) (2019). *Diseases of Swine.*

414. Sutherland MA, PJ Bryer, N Krebs, and JJ McGlone (2008). "Tail docking in pigs: acute physiological and behavioural responses." *Animal* 2(2): 292–297; Van Beirendonck S, B Driessen, G Verbeke, L Permentier, V Van de Perre, and R Geers (2012). "Improving survival, growth rate, and animal welfare in piglets by avoiding teeth shortening and tail docking." *Journal of Veterinary Behavior* 7(2): 88–93.

415. Schrey L, N Kemper, M Fels (2017). "Behaviour and skin injuries of sows kept in a novel group housing system during lactation." *Journal of Applied Animal Research,* 46(1): 749–757; Pedersen LJ (2017). "Overview of commercial production systems and their main welfare challenges." In

M Špinka (ed), *Advances in Pig Welfare*. Elsevier, pp 3–25; Baxter EM, IL Andersen, and SA Edwards (2017). "Sow welfare in the farrowing crate and alternatives." In M Špinka (ed), *Advances in Pig Welfare*. Elsevier, pp 27–72.

416. Chantziaras I, J Dewulf, T Van Limbergen, M Klinkenberg, and A Palzer (2018). "Factors associated with specific health, welfare and reproductive performance indicators in pig herds from five EU countries." *Preventive Veterinary Medicine*, 159: 106–114.

417. Jonas T (2015). "The vegetarian turned pig-farming butcher." In N Rose (ed), *Fair Food: Stories from a Movement Changing the World*. University of Queensland Press, St Lucia, Australia; Jonas T. (2015). "How to respond to vegan abolitionists." *Tammi Jonas: Food Ethics*. http://www.tammijonas. com/2015/03/24/how-to-respond-to-vegan-abolitionists/.

418. Spellberg B, GR Hansen, A Kar, CD Cordova, LB Price, and JR Johnson (2016). *Antibiotic Resistance in Humans and Animals*. National Academy of Medicine. Discussion Paper. https://nam.edu/antibiotic-resistance-in-humans-and-animals/; Wallace RG (2016). *Big Farms Make Big Flu: Dispatches on Infectious Disease, Agribusiness, and the Nature of Science*.

419. Spellberg B, et al. (2016). *Antibiotic Resistance in Humans and Animals*; Robinson TP, GRW Wint, G Conchedda, TP Van Boeckel, V Ercoli, et al. (2014). "Mapping the global distribution of livestock." *PLoS ONE*, 9(5):e96084; Mughini-Gras L, A Dorado-García, E van Duijkeren, G van den Bunt, CM Dierikx, et al. (2019). "Attributable sources of community-acquired carriage of *Escherichia coli* containing β-lactam antibiotic resistance genes: a population-based modelling study." *The Lancet Planetary Health*, 3: e357–e369.

420. Lambert ME and S D'Allaire (2009). "Biosecurity in swine production: Widespread concerns?" *Advances in Pork Production* 20: 139–148; Pitkin A, S Otake, and S Dee (2009). *Biosecurity Protocols for the Prevention of Spread of Porcine Reproductive and Respiratory Syndrome Virus*. Swine Disease Eradication Center, University of Minnesota College of Veterinary Medicine. https://datcp.wi.gov/Documents/PRRSVBiosecurityManual. pdf; Janni KA, LD Jacobson, SL Noll, CJ Cardona, HW Martin, and AE Neu (2016). "Engineering challenges and responses to the pathogenic avian influenza outbreak in Minnesota in 2015." 2016 American Society of Agricultural and Biological Engineers Annual International Meeting; Dewulf J and F Van Immerseel (2018). *Biosecurity in Animal Production and Veterinary Medicine: From Principles to Practice*. Acco, Leuven, Belgium.

421. Blanchette A (2015). "Herding species: Biosecurity, posthuman labor, and the American industrial pig." *Cultural Anthropology*, 30(4): 640–669; Wallace RG (2016). "Banksgiving." *Farming Pathogens*, 30 November. https://farmingpathogens.wordpress.com/2016/11/30/banksgiving/; Wickramage K and G Annunziata (2018). "Advancing health in migration governance, and migration in health governance." *The Lancet*, 392(10164):

2528–2530; Moyce SC and M Schenker (2018). "Migrant workers and their occupational health and safety." *Annual Review of Public Health* 39: 351–365.

422. Blanchette A (2015). "Herding species: Biosecurity, posthuman labor, and the American industrial pig"; Wallace RG (2016). "Banksgiving."

423. Wallace RG (2016). "Made in Minnesota."

424. Lee J, L Schulz, and G Tonsor (2019). "Swine producers' willingness to pay for Tier 1 disease risk mitigation under ambiguity." Selected Paper prepared for presentation at the 2019 Agricultural & Applied Economics Association Annual Meeting, Atlanta, GA, July 21 – July 23. https://ageconsearch.umn.edu/record/290908/files/Abstracts_19_05_13_19_46_48_31__129_186_252_93_0.pdf.

425. Wallace RG (2017). "Industrial production of poultry gives rise to deadly strains of bird flu H5Nx"; Briand FX, E Niqueux, A Schmitz, E Hirchaud, H Quenault, et al. (2018). "Emergence and multiple reassortments of French 2015–2016 highly pathogenic H5 avian influenza viruses." *Infection, Genetics and Evolution* 61: 208–214.

426. Belaich PC (2016). "Face à la grippe aviaire, les éleveurs du Sud-Ouest se remettent 'en ordre de marche'." *Le Monde*, 28 April. https://www.lemonde.fr/economie/article/2016/04/29/face-a-la-grippe-aviaire-les-eleveurs-du-sud-ouest-se-remettent-en-ordre-de-marche_4910900_3234.html.

427. Hill A (2015). "Moving from 'matters of fact' to 'matters of concern' in order to grow economic food futures in the Anthropocene." *Agriculture and Human Values* 32(3): 551-563; Wallace RG (2017). "Industrial production of poultry gives rise to deadly strains of bird flu H5Nx"; Maclean K, C Farbotko, and CJ Robinson (2019). "Who do growers trust? Engaging biosecurity knowledges to negotiate risk management in the north Queensland banana industry, Australia." *Journal of Rural Studies* 67: 101–110.

428. Collier SJ and A Lakoff (2008). "The problem of securing health." In A Lakoff and SJ Collier (eds), *Biosecurity Interventions: Global Health and Security in Question*. Columbia University Press, New York; Hinchliffe S (2013). "The insecurity of biosecurity: remaking emerging infectious diseases"; Allen J and S Lavau (2015). "'Just-in-time' disease: Biosecurity, poultry and power"; Gowdy J and P Baveye (2019). "An evolutionary perspective on industrial and sustainable agriculture." In G Lemaire, PCDF Carvalho, S Kronberg, and S Recous (eds), *Agroecosystem Diversity: Reconciling Contemporary Agriculture and Environmental Quality*. Academic Press, pp 425–433.

429. Akram-Lodhi AH (2015). "Land grabs, the agrarian question and the corporate food regime." *Canadian Food Studies* 2(2): 233–241; Montenegro de Wit M and A Iles (2016). "Toward thick legitimacy: creating a web of legitimacy for agroecology." *Elem. Sci. Anth.* 4: 115; Murray A (2018). "Meat cultures: Lab-grown meat and the politics of contamination." *BioSocieties* 13(2): 513–534; Wallace RG, K Okomoto, and A Liebman (2020). "Gated ecologies."

430. Powell J (2017). "Poultry farm sets up lasers to guard its organic hens from bird flu." *The Poultry Site*, 6 March. https://thepoultrysite.com/news/2017/03/poultry-farm-sets-up-lasers-to-guard-its-organic-hens-from-bird-flu; Benjamin M and S Yik (2019). "Precision livestock farming in swine welfare: A review for swine practitioners." *Animals* 9: 133; , Shen JH, C McDowell, Q Liu, M.Duff, et al. (2019). "Virus survival and fitness when multiple genotypes and subtypes of influenza A viruses exist and circulate in swine." *Virology* 532: 30–38.

431. Proudfoot C, S Lillico, and C Tait-Bukard (2019). "Genome editing for disease resistance in pigs and chickens." *Animal Frontiers* 9(3): 6–12.

432. Wallace RG (2016). "Cave/Man." In *Big Farms Make Big Flu: Dispatches on Infectious Disease, Agribusiness, and the Nature of Science*. Monthly Review Press, New York, pp 277–278.

433. Leonard C (2014). *The Meat Racket: The Secret Takeover of America's Food Business*; Wallace RG (2016). "Collateralized farmers"; Adams T, J-D Gerber, M Amacker, and T Haller. (2018). "Who gains from contract farming? Dependencies, power relations, and institutional change." *Journal of Peasant Studies* 46(7): 1435–1457.

434. Fracchia J (2017). "Organisms and objectifications: A historical-materialist inquiry into the 'Human and the Animal.'" *Monthly Review*, 68(10): 1–16; Wallace RG (2018). "Review of Paul Richards' *Ebola: How a People's Science Helped End an Epidemic*." (Zed Books, 2016)." https://antipodeonline.org/wp-content/uploads/2018/01/book-review_wallace-on-richards1.pdf.

435. Wallace RG (2016). "Made in Minnesota."

436. Moerman DE (1986) *Medicinal Plants of Native America*. Museum of Anthropology, Ann Arbor.

437. Ward SM, TM Webster, and LE Steckel (2013). "Palmer Amaranth (*Amaranthus palmeri*): A review." *Weed Technology* 27(1): 12–27; Chahal PS, JS Aulakh, M Jugulam, and AJ Jhala (2015). "Herbicide-resistant Palmer amaranth (*Amaranthus palmeri* S. Wats.) in the United States—mechanisms of resistance, impact, and management." In A Price, J Kelton, and L Sarunite (eds), *Herbicides: Agronomic Crops and Weed Biology*. InTech, Rijeka, Croatia, pp 1–29.

438. Ibid.

439. Webster TM and LM Sosnoskie (2010). "Loss of glyphosate efficacy: a changing weed spectrum in Georgia cotton." *Weed Science* 58(1): 73–79.

440. Kniss AR (2018). "Genetically engineered herbicide-resistant crops and herbicide-resistant weed evolution in the United States." *Weed Science* 66(2): 260–273.

441. Ward SM, TM Webster, and LE Steckel (2013). "Palmer Amaranth (*Amaranthus palmeri*): A review"; Chahal PS, JS Aulakh, M Jugulam, and AJ Jhala (2015). "Herbicide-resistant Palmer amaranth (*Amaranthus palmeri* S. Wats.) in the United States—mechanisms of resistance, impact, and management."

442. Price AJ, KS Balkcom, SA Culpepper, JA Kelton, RL Nichols, and H Schomberg (2011). "Glyphosate-resistant Palmer amaranth: a threat to conservation tillage." *J. Soil Water Conserv.* 66: 265–275; Menalled F, R Peterson, R Smith, W Curran, D Páez, and B Maxwell (2016). "The eco-evolutionary imperative: revisiting weed management in the midst of an herbicide resistance crisis." *Sustainability* 8(12): 1297.

443. Webster TM and LM Sosnoskie (2010). "Loss of glyphosate efficacy: a changing weed spectrum in Georgia cotton"; Ward SM, TM Webster, and LE Steckel (2013). "Palmer Amaranth (*Amaranthus palmeri*): A review"; Chahal PS, JS Aulakh, M Jugulam, and AJ Jhala (2015). "Herbicide-resistant Palmer amaranth (*Amaranthus palmeri* S. Wats.) in the United States—mechanisms of resistance, impact, and management."

444. Heap, I (2018).*The International Survey of Herbicide Resistant Weeds.* www.weedscience.com.

445. Ward SM, TM Webster, and LE Steckel (2013). "Palmer Amaranth (*Amaranthus palmeri*): A review"; Chahal PS, JS Aulakh, M Jugulam, and AJ Jhala (2015). "Herbicide-resistant Palmer amaranth (*Amaranthus palmeri* S. Wats.) in the United States—mechanisms of resistance, impact, and management."

446. Ibid. Culpepper AS, TM Webster, LM Sosnoskie, and AC York (2010). "Glyphosate-resistant Palmer amaranth in the US." In VK Nandula (ed), *Glyphosate Resistance: Evolution, Mechanisms, and Management.* J Wiley, Hoboken, NJ, pp 195–212; Culpepper AS, JS Richburg, AC York, LE Steckel, and LB Braxton (2011). "Managing glyphosate-resistant Palmer amaranth using 2,4-D systems in DHT cotton in GA, NC, and TN." *Proceedings of the 2011 Beltwide Cotton Conference.* National Cotton Council of America, Cordova, TN, p 1543; Price AJ, KS Balkcom, SA Culpepper, JA Kelton, RL Nichols, and H Schomberg (2011). "Glyphosate-resistant Palmer amaranth: a threat to conservation tillage"; Gaines TA, SM Ward, B Bekun, C Preston, JE Leach, and P Westra (2012). "Interspecific hybridization transfers a previously unknown glyphosate resistance mechanism in Amaranthus species." *Evol. Applic.* 5: 29–38.

447. Kumar V, R Liu, G Boyer, and PW Stahlman (2019). "Confirmation of 2, 4-D resistance and identification of multiple resistance in a Kansas Palmer amaranth (*Amaranthus palmeri*) population." *Pest Management Science.* https://doi.org/10.1002/ps.5400.

448. Alexander A (2017). "Court finds spraying of dicamba by third-party farmers an intervening cause [*Bader Farms, Inc. v. Monsanto Co.*]." National Agricultural Law Center. http://nationalaglawcenter.org/court-finds-spraying-dicamba-third-party-farmers-intervening-cause-bader-farms-inc-v-monsanto-co/; Hakim D (2017). "Monsanto's weed killer, dicamba, divides farmers." *New York Times*, 21 September. https://www.nytimes.com/2017/09/21/business/monsanto-dicamba-weed-killer.html; McCune M (2017). "A pesticide, a pigweed, and a farmer's

murder." *Planet Money*, National Public Radio, 14 June. https://www.npr. org/2017/06/14/532879755/a-pesticide-a-pigweed-and-a-farmers-murder.

449. Guthman J and S Brown (2016). "Whose life counts: Biopolitics and the 'Bright Line' of chloropicrin mitigation in California's strawberry industry." *Science, Technology, and Human Values* 41(3): 461–482; Moodie A (2017). "Fowl play: the chicken farmers being bullied by big poultry." *The Guardian*, 22 April. https://www.theguardian.com/sustainable-business/2017/ apr/22/chicken-farmers-big-poultry-rules.

450. Stanescu J (2013). "Beyond biopolitics: Animal studies, factory farms, and the advent of deading life." *PhaenEx*. 8(2): 135–160.

451. Larson BMH (2007). "Who's invading what? Systems thinking about invasive species." *Canadian Journal of Plant Science* 993–999; Biermann C and B Mansfield (2014). "Biodiversity, purity, and death: conservation biology as biopolitics." *Environment and Planning D: Society and Space* 32: 257–273; Barker K (2014). "Biosecurity: securing circulations from the microbe to the macrocosm." *The Geographical Journal* 181(4): 357–365; Larson BMH (2016). "New wine and old wineskins? Novel ecosystems and conceptual change." *Nature and Culture* 11(2): 148-164; Srinivasan K (2017). "Conservation biopolitics and the sustainability episteme." *Environment and Planning A* 49(7): 1458–1476.

452. Foucault M (2007). *Security, Territory, Population: Lectures at the Collège de France, 1977–78*. Palgrave Macmillan, Basingstoke.

453. Guthman J and S Brown (2016). "Whose life counts: Biopolitics and the 'Bright Line' of chloropicrin mitigation in California's strawberry industry."

454. Barker K (2014). "Biosecurity: securing circulations from the microbe to the macrocosm."

455. Davis AS and GB Frisvold (2017). "Are herbicides a once in a century method of weed control?" *Pest Manag Sci* 73: 2209–2220; Harker KN, C Mallory-Smith, BD Maxwell, DA Mortensen and RG Smith (2017). "Another view." *Weed Science* 65: 203–205.

456. Hindmarsh R (2005). "Green biopolitics and the molecular reordering of nature." Paper presented to the 'Mapping Biopolitics: Medical-Scientific Transformations and the Rise of New Forms of Governance' Workshop, European Consortium for Political Research Conference, Granada, Spain, 14–19 April 2005.

457. Harvey D (2001). "Globalization and the 'spatial fix.'" *Geographische Revue* 2: 23–30.

458. Harvey D (1982; 2006). *The Limits to Capital*; Walker R and M Stroper (1991). *The Capitalist Imperative: Territory, Technology and Industrial Growth*. Wiley; Wallace RG, L Bergmann, R Kock, M Gilbert, L Hogerwerf, R Wallace and M Holmberg (2015). "The dawn of Structural One Health: A new science tracking disease emergence along circuits of capital."

459. Stanescu J (2013). "Beyond biopolitics: Animal studies, factory farms, and the advent of deading life."

460. Clegg S (1989). *Frameworks of Power.* Sage, London.
461. Foucault M (2007). *Security, Territory, Population: Lectures at the College de France, 1977–78.*
462. Morton T (2013). *Hyperobjects: Philosophy and Ecology after the End of the World.* University of Minnesota Press, Minneapolis.
463. Chen GQ, YH He, S Qiang (2013). "Increasing seriousness of plant invasions in croplands of Eastern China in relation to changing farming practices: A case study." *PLoS ONE,* 8(9): e74136; Robinson TP, GRW Wint, G Conchedda, TP Van Boeckel, V Ercoli, et al. (2014). "Mapping the global distribution of livestock."
464. Despommier D, BR Ellis, and BA Wilcox (2006) "The role of ecotones in emerging infectious diseases." *Ecohealth* 3(4): 281–289; Wallace RG (2016). *Big Farms Make Big Flu: Dispatches on Infectious Disease, Agribusiness, and the Nature of Science.*
465. Lewontin R (1998). "The maturing of capitalist agriculture: Farmer as proletarian"; Coppin D (2003). "Foucauldian hog futures: The birth of mega-hog farms." *The Sociological Quarterly* 44(4): 597–616; Baird IG (2011). "Turning land into capital, turning people into labour: Primitive accumulation and the arrival of large-scale economic land concessions in the Lao People's Democratic Republic." *New Proposals: Journal of Marxism and Interdisciplinary Inquiry* 5: 10–26.
466. Rotz S and EDG. Fraser (2015). "Resilience and the industrial food system: analyzing the impacts of agricultural industrialization on food system vulnerability." *J Environ Stud Sci.* 5: 459–473; Wallace RG, R Alders, R Kock, T Jonas, R Wallace, and L Hogerwerf (2019). "Health before medicine: Community resilience in food landscapes."
467. Franklin HB (1979). "What are we to make of J.G. Ballard's *Apocalypse?*" In TD Calreson (ed), *Voices for the Future: Essays on Major Science Fiction Writers, Volume Two.* Bowling Green State University Popular Press, Bowling Green, OH.
468. Hincliffe S (2013). "The insecurity of biosecurity: remaking emerging infectious diseases"; Karatani K (2014). *The Structure of World History: From Modes of Production to Modes of Exchange.* Duke University Press, Durham, NC; Wallace RG, L Bergmann, R Kock, M Gilbert, L Hogerwerf, R Wallace and M Holmberg (2015). "The dawn of Structural One Health: A new science tracking disease emergence along circuits of capital"; Foster JB and P Burkett (2016). *Marx and the Earth: An Anti-Critique.* Brill Academic Publishers, The Netherlands; Harvey D (2018). "Marx's refusal of the labour theory of value: Reading Marx's *Capital* with David Harvey"; Kallis G and E Swyngedouw (2018). "Do bees produce value? A conversation between an ecological economist and a Marxist geographer." *Capitalism Nature Socialism* 29(3): 36–50; Wallace RG, R Alders, R Kock, T Jonas, R Wallace, and L Hogerwerf (2019). "Health before medicine: Community resilience in food landscapes."

469. Keck F (2019). "Livestock Revolution and ghostly apparitions: South China as a sentinel territory for influenza pandemics." *Current Anthropology* 60(S20):S251–S259; Tsing A (2017). "The buck, the bull, and the dream of the stag: Some unexpected weeds of the Anthropocene." *Soumen Anthropologi* 42(1): 3–21; Tsing A, J Deger, AK Saxena, and E Gan (eds). (2020). *Feral Atlas: The More-than-Human Anthropocene.* Stanford University Press, Palo Alto, CA.

470. Wallace R and RG Wallace (2015). "Blowback: new formal perspectives on agriculturally-driven pathogen evolution and spread"; Gowdy J and P Baveye (2019). "An evolutionary perspective on industrial and sustainable agriculture." In G Lemaire, PCDF Carvalho, S Kronberg, and S Recous (eds), *Agroecosystem Diversity: Reconciling Contemporary Agriculture and Environmental Quality.*

471. Meiners SJ, STA Pickett, and ML Cadenasso (2002). "Exotic plant invasions over 40 years of old field successions: community patterns and associations." *Ecography* 25: 215–223; Jordan NR, L Aldrich-Wolfe, SC Huerd, D Larson, and G Muehlbauer (2012). "Soil–occupancy effects of invasive and native grassland plant species on composition and diversity of mycorrhizal associations." *Invasive Plant Science and Management* 5(4): 494–505; Stein S (2020). "Witchweed and the ghost: A parasitic plant devastates peasant crops on capital-abandoned plantations in Mozambique." In A Tsing, J Deger, AK Saxena, and E Gan (eds), *Feral Atlas: The More-than-Human Anthropocene.* Stanford University Press, Palo Alto, CA.

472. Simberloff D (2012). "Nature, natives, nativism, and management: Worldviews underlying controversies in invasion biology." *Environmental Ethics* 34(1): 5–25; Code L (2013). "Doubt and denial: Epistemic responsibility meets climate change skepticism." *Onati Socio-Legal Series* 3(5): 838853; Doan MD (2016). "Responsibility for collective inaction and the knowledge condition." *Social Epistemology* 30: 532–54.

473. Levins R (1998). "The internal and external in explanatory theories." *Science as Culture* 7(4): 557–582; Levins R (2006). "Strategies of abstraction." *Biol Philos* 21: 741–755; Winther RG (2006). "On the dangers of making scientific models ontologically independent: taking Richard Levins' warnings seriously." *Biol Philos* 21: 703–724; Schizas D (2012). "Systems ecology reloaded: A critical assessment focusing on the relations between science and ideology." In GP Stamou (ed), *Populations, Biocommunities, Ecosystems: A Review of Controversies in Ecological Thinking.* Bentham Science Publishers, Sharjah; Nikisianis N and GP Stamou (2016). "Harmony as ideology: Questioning the diversity-stability hypothesis." *Acta Biotheoretica* 64(1): 33–64.

474. Wallace R, LF Chaves, LR Bergmann, C Ayres, L Hogerwerf, R Kock, and RG Wallace (2018). *Clear-Cutting Disease Control: Capital-Led Deforestation, Public Health Austerity, and Vector-Borne Infection.*

475. Farmer P (2008). "Challenging orthodoxies: the road ahead for health

and human rights." *Health Hum. Rights* 10(1): 519; Sparke M (2009). "Unpacking economism and remapping the terrain of global health." In A Kay, OD Williams (eds), *Global Health Governance: Crisis, Institutions and Political Economy.* Springer, Cham; Chiriboga D, P Buss, AE Birn, J Garay, C Muntaner, and L Nervi (2015). "Investing in health." *The Lancet* 383(9921): 949; Sparke M (2017). "Austerity and the embodiment of neo-liberalism as ill-health: Towards a theory of biological sub-citizenship." *Social Science Medicine* 187: 287–295.

476. Chapura M (2007). *Actor Networks, Economic Imperatives and the Heterogeneous Geography of the Contemporary Poultry Industry.* MA thesis, Department of Geography, University of Georgia; Food and Water Watch (2012). *Public Research, Private Gain: Corporate Influence Over University Agricultural Research.* https://www.foodandwaterwatch.org/sites/default/files/Public%20Research%20Private%20Gain%20Report%20April%202012.pdf; Pardy PG, JM Alston, C Chang-Kang, TM Hurley, RS Andrade, et al. (2018). "The shifting structure of agricultural R&D: Worldwide investment patterns and payoffs." In N Kalaitzandonakes, EG Carayannis, E Grigoroudis, and S Rozakis (2018). *From Agriscience to Agribusiness: Theories, Policies and Practices in Technology Transfer and Commercialization.* Springer, Cham.

477. Hulme PE (2009). "Trade, transport and trouble: managing invasive species pathways in an era of globalization." *Journal of Applied Ecology* 46: 10–18; Blanchette A (2015). "Herding species: Biosecurity, posthuman labor, and the American industrial pig"; Bagnato A (2017). "Microscopic colonialism." *E-Flux.* https://www.e-flux.com/architecture/positions/153900/microscopic-colonialism/.

478. Lezaun J and N Porter (2015). "Containment and competition: transgenic animals in the One Health agenda." *Soc Sci Med* 129: 96–105; Wallace RG (2016). "Protecting H3N2v's privacy." In *Big Farms Make Big Flu: Dispatches on Infectious Disease, Agribusiness, and the Nature of Science.* Monthly Review Press, New York, pp 319–321; Borkenhagen LK, MD Salman, M Mai-Juan, GC Gray (2019). "Animal influenza virus infections in humans: A commentary." *International Journal of Infectious Diseases,* 88: 113–119; Gorsich EE, RS Miller, HM Mask, C Hallman, K Portacci, and CT Webb (2019). "Spatio-temporal patterns and characteristics of swine shipments in the U.S. based on Interstate Certificates of Veterinary Inspection"; Okamoto K, A Liebman, and RG Wallace (2020). "At what geographic scales does agricultural alienation amplify foodborne disease outbreaks? A statistical test for 25 U.S. states, 1970–2000."

479. Norgaard RB (1984). "Coevolutionary agricultural development." *Economic Development and Cultural Change,* 32(3): 525–546; Noailly J (2008). "Coevolution of economic and ecological systems. An application to agricultural pesticide resistance." *Journal of Evolutionary Economics* 18: 1–29; Tsing A (2009). "Supply chains and the human

condition"; Moreno-Peñaranda R and G Kallis (2010). "A coevolutionary understanding of agroenvironmental change: A case-study of a rural community in Brazil." *Ecological Economics* 69(4): 770–778; Coq-Huelva D, A Higuchi, R Alfalla-Luque, R Burgos-Morán, and R Arias-Gutiérrez (2017). "Co-evolution and bio-social construction: The Kichwa agroforestry systems (*Chakras*) in the Ecuadorian Amazonia." *Sustainability* 9(10): 1920; Giraldo OF (2019). *Political Ecology of Agriculture: Agroecology and Post-Development.*

480. IPES-Food (2016). *From Uniformity to Diversity: A Paradigm Shift from Industrial Agriculture to Diversified Agroecological Systems.* Louvain-la-Neuve, Belgium. http://www.ipes-food.org/_img/upload/files/UniformityToDiversity_FULL. pdf; IPES-Food (2018). *Breaking Away from Industrial Food and Farming Systems: Seven Case Studies of Agroecological Transition.* Louvain-la-Neuve, Belgium. http://www.ipes-food.org/_img/upload/files/CS2_web.pdf; Chappell MJ (2018). *Beginning to End Hunger: Food and the Environment in Belo Horizonte, Brazil, and Beyond;* Arias PF, T Jonas, and K Munksgaard (eds) (2019). *Farming Democracy: Radically Transforming the Food System from the Ground Up.* Australian Food Sovereignty Alliance; Vivero-Pol JL, T Ferrando, O De Schutter, and U Mattei (eds) (2019). *Routledge Handbook of Food as a Commons.* Routledge, New York.

481. Liebman M, ER Gallandt, and LE Jackson (1997). "Many little hammers: ecological management of crop-weed interactions." *Ecology in Agriculture* 1: 291–343; Enticott G (2008). "The spaces of biosecurity: prescribing and negotiating solutions to bovine tuberculosis." *Environment and Planning A* 40: 1568–1582; Wallace RG (2016). "A probiotic ecology." In *Big Farms Make Big Flu: Dispatches on Infectious Disease, Agribusiness, and the Nature of Science.* Monthly Review Press, New York, pp 250–256; Midega CA, JO Pittchar, JA Pickett, GW Hailu, and ZR Khan (2018). "A climate-adapted push-pull system effectively controls fall armyworm, *Spodoptera frugiperda* (J E Smith), in maize in East Africa." *Crop Protection* 105: 10–15; Wallace RG, R Alders, R Kock, T Jonas, R Wallace, and L Hogerwerf (2019). "Health before medicine: Community resilience in food landscapes."

482. Holt-Giménez E and M Altieri (2016). *Agroecology "Lite:" Cooptation and Resistance in the Global North.* Food First—Institute for Food and Development Policy. https://foodfirst.org/agroecology-lite-cooptation-and-resistance-in-the-global-north/; Giraldo OF (2019). *Political Ecology of Agriculture: Agroecology and Post-Development.*

Pandemic Research for the People

483. Agroecology and Rural Economics Research Corps (2020). "Pandemic Research for the People." ARERC, 25 March. https://arerc.wordpress.com/pandemic-research-for-the-people/; Pandemic Research for the People (2020). "About PReP." https://www.prepthepeople.net/about.

The Bright Bulbs

484. Luce E (2020). "Inside Trump's coronavirus meltdown." *Financial Times*, 14 May. https://www.ft.com/content/97dc7de6-940b-11ea-abcd-371e24b679ed.

485. Scott K (2002). TikTok video, 14 April. Available online at https://www.tiktok.com/@kyscottt/video/6815758866769349894.

486. Wilson J (2020). "US lockdown protests may have spread virus widely, cellphone data suggests." *The Guardian*, 18 May. https://www.theguardian.com/us-news/2020/may/18/lockdown-protests-spread-coronavirus -cellphone-data.

487. Spinney L (2020). "The coronavirus slayer! How Kerala's rock star health minister helped save it from Covid-19." *The Guardian*, 14 May. https://www.theguardian.com/world/2020/may/14/the-coronavirus-slayer-how-keralas-rock-star-health-minister-helped-save-it-from-covid-19; Kretzschmar ME, G Rozhnova, MCJ Bootsma, M van Boven, JHHM van de Wijgert, MJM Bonten (2020). "Impact of delays on effectiveness of contact tracing strategies for COVID-19: a modelling study." *The Lancet Public Health*, 16 July. https://doi.org/10.1016/S2468-2667(20)30157-2; Luo E, N Chong, C Erikson, C Chen, S Westergaard, E Salsberg, and P Pittman (2020). "Contact tracing workforce estimator." Fitzhugh Mullan Institute for Health Workforce Equity, George Washington University. https://www.gwhwi.org/estimator-613404.html.

488. Pan A, et al. (2020). "Association of public health interventions with the epidemiology of the COVID-19 outbreak in Wuhan, China." *JAMA* 323(19): 1915–1923; Black G (2020). "Vietnam may have the most effective response to Covid-19." *The Nation*, 24 April. https://www.thenation.com/article/world/coronavirus-vietnam-quarantine-mobilization/.

489. Gurba M (2020). "Be about it: a history of mutual aid has prepared POC for this moment." *REMEZCLA*, 14 May https://remezcla.com/features/culture/southern-solidarity-mutual-aid-history-and-coronavirus; Derysh I (2020). "States smuggle COVID-19 medical supplies to avoid federal seizures as House probes Jared Kushner." *Salon*, 21 April https://www.salon.com/2020/04/21/states-smuggle-covid-19-medical-supplies-to-avoid-federal-seizures-as-house-probes-jared-kushner; Artenstein AW (2020). "In pursuit of PPE." *New England Journal of Medicine* 382:e46.

490. Richert C (2020). "Minnesota counties say contact tracing is taking too long." *MPRNews*, 11 May. https://www.mprnews.org/story/2020/05/11/minnesota-counties-say-contact-tracing-is-taking-too-long.

491. Goodman JD, WK Rashbaum, and JC Mays (2020). "DeBlasio strips control of virus tracing from health department." *New York Times*, 7 May. https://www.nytimes.com/2020/05/07/nyregion/coronavirus-contact-tracing-nyc.html; Sanders A (2020). "NYC coronavirus contract tracers program turns disastrous after hiring too many remote workers." *New York Daily News*, 28 May. https://www.nydailynews.com/coronavirus/ny-coronavirus-nyc-contact-tracer-hiring-hospital-system-remote-20200528-xpdt4pefkrgwhiso73cvgvj4ge-story.html.

492. Trump D (2020). "Donald Trump speech transcript at PA distribution center for coronavirus relief supplies." *Rev*, 14 May. https://www.rev.com/blog/transcripts/donald-trump-speech-transcript-at-pennsylvania-distribution-center-for-coronavirus-relief-supplies; Our World in Data (2020). "Daily COVID-19 tests per thousand people." https://ourworldindata.org/grapher/full-list-daily-covid-19-tests-per-thousand.

493. Romm T, J Stein, and E Werner (2020). "2.4 million Americans filled job claims last week, bringing nine-week total to 38.6 million." *Washington Post*, 21 May. https://www.washingtonpost.com/business/2020/05/21/unemployment-claims-coronavirus/.

494. Táíwò O (2020). "Corporations are salivating over the coronavirus pandemic." *The New Republic*, 3 April. https://newrepublic.com/article/157159/corporations-salivating-coronavirus-pandemic; Picchi A (2020). "Trump adviser says America's 'human capital stock' ready to return to work, sparking anger." CBS News, 26 May. https://www.cbsnews.com/news/human-capital-stock-kevin-hassett-trump-economic-advisor-back-to-work; Hartman M (2020). "Half of Americans who lost work or wages are getting $0 jobless benefits." *Marketplace*, 18 May. https://www.marketplace.org/2020/05/18/half-of-americans-who-lost-work-or-wages-are-getting-0-jobless-benefits/; Brenner R (2020). "Escalating plunder." *New Left Review* 123: 5–22.

495. Glenza J (2020). "Up to 43m Americans could lose health insurance amid pandemic, report says." *The Guardian*, 10 May. https://www.theguardian.com/us-news/2020/may/10/us-health-insurance-layoffs-coronavirus.

496. BBC News (2020). "Wilbur Ross says coronavirus could boost US jobs." *BBC News*, 31 January. https://www.bbc.com/news/business-51276323.

497. Chang GC (2020). "Many smart people, knowing that #China would dominate the world . . ." Twitter, 19 February. https://twitter.com/GordonGChang/status/1230147427795185667.

498. Associated Press (2020). "US Senator criticized for telling students China is to blame for Covid-19." *The Guardian*, 17 May. https://www.theguardian.com/us-news/2020/may/17/senator-ben-sasse-china-coronavirus-graduation-speech; Rogin J (2020). "The coronavirus crisis is turning Americans in both parties against China." *Washington Post*, 8 April. https://www.washingtonpost.com/opinions/2020/04/08/coronavirus-crisis-is-turning-americans-both-parties-against-china/.

499. Chiu A (2020). "Trump has no qualms about calling coronavirus the 'Chinese Virus.' That's a dangerous attitude, experts say." *Washington Post*, 20 March. https://www.washingtonpost.com/nation/2020/03/20/coronavirus-trump-chinese-virus; Day M (2020). "No act of God." *Jacobin*, 19 May. https://www.jacobinmag.com/2020/05/no-act-of-god.

500. Fisher M (2020). "Coronavirus 'hits all the hot buttons' for how we misjudge risk." *New York Times*, 13 February. https://www.nytimes.com/2020/02/13/world/asia/coronavirus-risk-interpreter.html.

501. Cornwall W (2020). "Crushing coronavirus means 'breaking the habits of a lifetime.' Behavior scientists have some tips." *Science*, 16 April. https://www.sciencemag.org/news/2020/04/crushing-coronavirus-means-breaking-habits-lifetime-behavior-scientists-have-some-tips.

502. WABC (2020). "Coronavirus news: NYC poison control sees uptick in Lysol, bleach exposures after Trump's comments on disinfectants." Eyewitness News ABC7, 25 April. https://abc7ny.com/lysol-bleach-president-trump-nyc-health/6128990/.

503. The Brian Lehrer Show (2017). "When Ivanka Trump and Senator Schumer share cocktails in the Hamptons." WNYC website, 10 July. https://www.wnyc.org/story/when-ivanka-trump-and-sen-schumer-share-cocktails-hamptons/.

504. Vogell H and K Sullivan (2020). "Trump's company paid bribes to reduce property taxes, assessors say." ProPublica, 11 March. https://www.propublica.org/article/trumps-company-paid-bribes-to-reduce-property-taxes-assessors-say.

505. Gindin S and L Panitch (2012). *The Making of Global Capitalism*. Verso: New York.

506. Tharoor I (2020). "Trump's pandemic response underscores the crisis in global politics." *Washington Post*, 17 April. https://www.washingtonpost.com/world/2020/04/17/trumps-pandemic-response-underscores-crisis-global-politicst/.

507. Karni A and M Haberman (2020). "Trump announces his 'opening the country' council." *The New York Times*, 14 April. https://www.nytimes.com/2020/04/14/us/politics/coronavirus-trump-reopening-council.html.

508. Oprysko C, B Ehley, R Roubein, and Q Forgey (2020). "Trump kicks off a day of whiplash over future of coronavirus task force." *Politico*, 6 May. https://www.politico.com/news/2020/05/06/trump-white-house-coronavirus-task-force-239900; Gangitano A (2020). "Trump uses Defense Production Act to order meat processing plants to stay open." *The Hill*, 28 April https://thehill.com/homenews/administration/495175-trump-uses-defense-production-act-to-order-meat-processing-plants-to; Kwiatkowski M and TL Nadolny (2020). "'It makes no sense': Feds consider relaxing infection control in US nursing homes." *USA Today*, 4 May. https://www.usatoday.com/story/news/investigations/2020/05/04/coronavirus-nursing-homes-feds-consider-relaxing-infection-control/3070288001.

509. Schwartz M (2020). "Governors divide by party on Trump plan to reopen businesses shut by coronavirus." NPR, 17 April. https://www.npr.org/2020/04/17/837579713/governors-divide-by-party-on-trump-plan-to-reopen-businesses-shut-by-coronavirus.

510. James F (2009). "Call 'swine flu' H1N1 instead: Ag Sec'y." NPR, 28 April. https://www.npr.org/sections/health-shots/2009/04/call_swine_flu_h1n1_instead_ob.html; Farm Lands of Guinea Limited (2011). "Farm Lands of Guinea completes reverse merger and investment valuing the company at USD$45 million." *Cision PR Newswire*, 4 March. https://www.prnewswire.com/news-releases/farm-lands-of-guinea

-completes-reverse-merger-and-investment-valuing-the-company-at-usd45-million-117415048.html.

511. Toosi N (2020). "Biden ad exposes a rift over China on the left." *Politico*, 23 April. https://www.politico.com/news/2020/04/23/biden-ad-exposes-left-rift-china-202241; Biden J (2020). "Joe Biden: My plan to safely reopen America." *New York Times*, 12 April. https://www.nytimes.com/2020/04/12/opinion/joe-biden-coronavirus-reopen-america.html.

512. Oprysko C and M Caputo. (2020). "Trump, Biden speak by phone about coronavirus response." *Politico*, 6 April https://www.politico.com/news/2020/04/06/trump-biden-dnc-convention-168323.

513. Cohen E (2020). "China says coronavirus can spread before symptoms show—calling into question US containment strategy." CNN, 26 January. https://www.cnn.com/2020/01/26/health/coronavirus-spread-symptoms-chinese-officials/index.html.

514. Chan JFW, et al. (2020). "A familial cluster of pneumonia associated with the 2019 novel coronavirus indicating person-to-person transmission: a study of a family cluster." *The Lancet*, 24 January. https://www.thelancet.com/journals/lancet/article/PIIS0140-6736(20)30154-9/fulltext.

515. Saturday Night Live (2020). "Dr. Anthony Fauci Cold Open—SNL." YouTube, 25 April. https://www.youtube.com/watch?v=uW56CL0pk0g; Johnson, LM (2020). "Doughnuts featuring Dr. Fauci's face are quickly becoming a nationwide hit." CNN, 26 March; https://www.cnn.com/2020/03/26/us/dr-fauci-doughnuts-trnd/index.html.

516. C-SPAN (2020). "Fauci on NSC global health office." C-SPAN, 19 March. https://www.c-span.org/video/?c4862190/user-clip-fauci-nsc-global-health-office.

517. Rev.com (2020). "NIH director to testify before House on coronavirus response as cases grow in U.S." Rev website, 11 March. https://www.rev.com/transcript-editor/shared/opVK9vJZvEC1TLj1TmE5nDfmsB2E-SUETmEnVbgr7N_QZUfxRTZMwEpZMuDZ_PR7mM4b7sV3qBOMpt9pNsAvYcKuSjDI?loadFrom=PastedDeeplink&ts=3097blka.

518. Matthews GJ (2018) "Family caregivers, AIDS narratives, and the semiotics of the bedside in Colm Tóibín's *The Blackwater Lightship*." *Critique* 60(3): 289–299.

519. Price JR (2017). "The treatment and prevention of HIV bodies." In M Brettschneider, S Burgess, and C Keating (eds), *LGBTQ Politics: A Critical Reader*. New York University Press, New York, pp 54–71.

520. Hoppe T (2018). *Punishing Disease: HIV and the Criminalization of Sickness*. University of California Press, Oakland.

521. Fauci AS, HC Lane, and RR Redfield (2020). "Covid-19—navigating the uncharted." *New England Journal of Medicine* 382: 1268–1269.

522. Faust JS and C del Rio (2020). "Assessment of deaths from COVID-19 and from seasonal influenza." *JAMA Internal Medicine*, 14 May. https://jama-network.com/journals/jamainternalmedicine/fullarticle/2766121.

523. Smith KF, et al. (2014). "Global rise in human infectious disease outbreaks." *Journal of the Royal Society* 11(101): 20140950.

524. Médecins Sans Frontières (2020). "DRC Ebola outbreaks: crisis update —May 2020." Médecins Sans Frontières, 18 May. https://www.msf.org/drc-ebola-outbreak-crisis-update.

525. Ota M (2020). "Will we see protection or reinfection in COVID-19?" *Nature Reviews Immunology* 20(6): 351; Harding L (2020). "'Weird as hell': the Covid-19 patients who have symptoms for months." *The Guardian*, 15 May. https://www.theguardian.com/world/2020/may/15/weird-hell-professor-advent-calendar-covid-19-symptoms-paul-garner; Parshley L (2020). "The emerging long-term complications of Covid-19, explained." *Vox*, 8 May. https://www.vox.com/2020/5/8/21251899/coronavirus-long-term-effects-symptoms.

526. Kreston R (2013). "The public health legacy of the 1976 swine flu outbreak." *Discover*, 30 September. https://www.discovermagazine.com/health/the-public-health-legacy-of-the-1976-swine-flu-outbreak.

527. Reinhard B, E Brown, and N Satjia (2020). "Trump says he can bring in coronavirus experts quickly. The experts say it is not that simple." *Washington Post*, 27 February https://www.washingtonpost.com/investigations/trump-says-he-can-bring-in-coronavirus-experts-quickly-the-experts-say-it-is-not-that-simple/2020/02/27/6ce214a6-5983-11ea-8753-73d96000faae_story.html.

528. Wallace RG (2010). "The Alan Greenspan strain." *Farming Pathogens* blog, 30 March. https://farmingpathogens.wordpress.com/2010/03/30/the-alan-greenspan-strain/.

529. St Clair J (2020). "Roaming charges: bernt offerings." *Counterpunch*, 17 April. https://www.counterpunch.org/2020/04/17/roaming-charges-bernt-offerings/.

530. Bendavid E, et al. (2020). "COVID-19 antibody seroprevalence in Santa Clara County, California." *MedRvix*, 30 April. https://www.medRxiv.org/content/10.1101/2020.04.14.20062463v2.

531. Lee SM (2020). "JetBlue's founder helped fund a Stanford study that said the coronavirus wasn't that deadly." *BuzzFeed News*, 15 May. https://www.buzzfeednews.com/article/stephaniemlee/stanford-coronavirus-neeleman-ioannidis-whistleblower.

532. Jasper C, C Ryan, and A Kotoky (2020). "An $85 billion airline rescue may only prolong the pain." *Bloomberg*, 2 May. https://www.bloomberg.com/news/articles/2020-05-02/coronavirus-airline-bailouts-a-guide-to-85-billion-in-state-aid.

533. Dutkiewicz J, A Taylor, and T Vettese (2020). "The Covid-19 pandemic shows we must transform the global food system." *The Guardian*, 16 April. https://www.theguardian.com/commentisfree/2020/apr/16/coronavirus-covid-19-pandemic -food-animals.

534. Wallace RG, A Liebman, LF Chaves, and R Wallace (2020). "COVID-19 and circuits of capital." This volume.

535. Huber M (2019). "Ecological politics for the working class." *Catalyst* (3)1. https://catalyst-journal.com/vol3/no1/ecological-politics-for-the-working-class.

536. Wurgaft BA (2019). *Meat Planet*. University of California Press, Oakland; Ellis EG (2019). "I'm a vegetarian—will I eat lab-grown meat?" *Wired*, 27 November. https://www.wired.com/story/vegetarian-ethics-lab-grown-meat/.

537. Murray A (2018). "Meat cultures: Lab-grown meat and the politics of contamination." *BioSocieties* 13: 513–534.

538. Puig N (2014). *Bédouins sédentarisés et société citadine à Tozeur*. Karthala, Paris.

539. Weizman E (2015). *The Conflict Shoreline: Colonization as Climate Change in the Negev Desert*. Steidel, Göttingen.

540. Amuasi, JH et al (2020). "Calling for a COVID-19 One Health research coalition." *The Lancet* 395(10236): 1543–1544.

541. Shah S and A Goodman (2020). "Sonia Shah: 'It's time to tell a new story about coronavirus—our lives depend on it.'" *Democracy Now!*, 17 July. https://www.democracynow.org/2020/7/17/sonia_shah_its_time_to_tell.

542. Wallace RG, et al. (2015). "The dawn of Structural One Health: a new science tracking disease emergence along circuits of capital." *Social Science and Medicine* 129: 68–77.

543. Johnson W (2017). *River of Dark Dreams: Slavery and Empire in the Cotton Kingdom*. Harvard University Press, Cambridge.

544. Luna AG, B Ferguson, O Giraldo, B Schmook, and EMA Maya (2019). "Agroecology and restoration ecology: fertile ground for Mexican peasant territoriality?" *Agroecology and Sustainable Food Systems* 43(10): 1174–1200.

545. Davis D (1996). "Gender, indigenous knowledge, and pastoral resource use in Morocco." *Geographical Review* 86(2): 284–288.

546. Jacobs R (2017). "An urban proletariat with peasant characteristics: land occupations and livestock raising in South Africa." *Journal of Peasant Studies* 45(5-6): 884–903.

547. Allred BW, SD Fuhlendorf, and RG Hamilton (2011). "The role of herbivores in Great Plains conservation: comparative ecology of bison and cattle." *Ecosphere* 2(3): 1–17.

548. Sharma D (2017). *Technopolitics, Agrarian Work and Resistance in Post-Green Revolution Indian Punjab*. Ph.D Dissertation available at https://ecommons.cornell.edu/handle/1813/59067.

549. Augustine D, A Davidson, K Dickinson, and B Van Pelt (2019). "Thinking like a grassland: Challenges and opportunities for biodiversity conservation in the Great Plains of North America." *Rangeland Ecology & Management*. https://www.sciencedirect.com/science/article/pii/S1550742419300697.

550. Gerber P, I Touré, A Ickowicz, I Garba, and B Toutain (2012). *Atlas of Trends in Pastoral Systems in the Sahel*. FAO & CIRAD, Rome.

551. Vettese T (2019). "The political economy of half-earth." *The Bullet*, 30 January. https://socialistproject.ca/2019/01/the-political-economy-of-half-earth/.

552. Cohen DA (2020). ""The problem isn't some people's taste for seemingly strange delicacies . . ." Twitter, 16 April. https://twitter.com/aldatweets/status/1250883909145026573.

553. Schmitz O (2016). "How 'natural geoengineering' can help slow global warming." *Yale Environment 360*, 25 January. https://e360.yale.edu/features/how_natural_geo-engineering_can_help_slow_global_warming.

554. Liang J, T Reynolds, A Wassie, C Collins, and A Wubalem (2016). "Effects of exotic *Eucalyptus spp.* plantationson soil properties in and around sacred natural sites in the northern Ethiopian Highlands." *AIMS Agriculture and Food* 1(2): 175–193.

555. Soto-Shoender JR, RA McCleery, A Monadjem, and DC Gwinn (2018). "The importance of grass cover for mammalian diversity and habitat associations in a bush encroached savanna." *Biological Conservation* 221:127–136.

556. Tishkov AA (2010). "Fires in steppes and savannas." In VM Kotlyanov (ed), *Natural Disasters—Volume II. Encyclopedia of Life Support Systems*, UK, pp 144–158.

557. Lynch J and R Pierrehumbert (2019). "Climate impacts of cultured meat and beef cattle." *Sustainable Food Systems*, 19 February https://www.frontiersin.org/articles/10.3389/fsufs.2019.00005/full.

558. Thieme R (2017). "The gruesome truth about lab-grown meat." *Slate*, 11 July. https://slate.com/technology/2017/07/the-gruesome-truth-about-lab-grown-meat.html.

559. Anonymous (2019). "Our meatless future: How the $1.8T global meat market gets disrupted." *CB Insights*, 13 November. https://www.cbinsights.com/research/future-of-meat-industrial-farming/.

560. Wallace RG (2019). "Redwashing capital: Left tech bros are honing Marx into a capitalist tool." *Uneven Earth*, 11 July. http://unevenearth.org/2019/07/redwashing-capital/.

561. Edelman M (2019). "Hollowed out Heartland, USA: How capital sacrificed communities and paved the way for authoritarian populism." *Journal of Rural Studies*, 10 November. https://www.sciencedirect.com/science/article/pii/S0743016719305157.

562. Anonymous (2007). "Wilderswil declaration on livestock diversity." La Via Campesina, 11 September. https://viacampesina.org/en/wilderswil-declaration -on-livestock-diversity/.

563. Swagemakers P, MDD Garcia, AO Torres, H Oostindie, and JCJ Groot (2017). "A values-based approach to exploring synergies between livestock farming and landscape conservation in Galicia (Spain)." *Sustainability* 9(11), 1987. https://www.mdpi.com/2071-1050/9/11/1987/htm.

564. Qualman D (2017). "Agribusiness takes all: 90 years of Canadian net farm income." *Darrin Qualman* blog, 28 February. https://www.darrinqualman.com/canadian-net-farm-income/.

565. CLOC—Via Campesina Secretary (2020). *CLOC—Via Campesina: Returning to the Countryside*. La Via Campesina, 14 April. https://

viacampesina.org/en/cloc-via-campesina-returning-to-the
-countryside/.

566. Mann G (2020). "Irrational expectations." *Viewpoint Magazine*, 29 April. https://www.viewpointmag.com/2020/04/29/irrational-expectations/.

567. Ajl M (2019). "How much will the US Way of Life © have to change?" *Uneven Earth*, 10 June. http://unevenearth.org/2019/06/how-much-will -the-us-way-of-life-have-to-change/.

568. Huber M (2020). "Socialize the food system." *Tribune*, 19 April. https:// www.tribunemag.co.uk/2020/04/socialise-the-food-system.

569. Perfecto I and J Vandermeer (2010). "The agroecological matrix as alternative to the land-sparing/agriculture intensification model." *PNAS* 107(13) 5786–5791; Smith A (2020). "To combat pandemics, intensify agriculture." The Breakthrough Institute, 13 April. https://thebreakthrough.org/issues/ food/zoonosis.

570. McDonald JM (2020). *Consolidation in US Agriculture Continues*. United States Department of Agriculture Economics Research Service, 3 February. https://www.ers.usda.gov/amber-waves/2020/february/consolidation-in -us-agriculture-continues/.

571. Marris E (2020). "GMOs are an ally in a changing climate." *Wired*, 1 April. https:// www.wired.com/story/future-food-will-need-gmo-organic-hybrid/.

572. Patel R and J Goodman (2020). "The Long New Deal." *Journal of Peasant Studies* 47(3): 431–463.

573. Kampf-Lassin M (2019). "A Popeyes chicken sandwich under socialism." *Jacobin*, 27 August. https://www.jacobinmag.com/2019/08/popeyes -chicken-sandwich-fast-food-workers.

574. Esteva G, DIG Luna, I Ragazzini (2014). "Mandar obedeciendo en territorio Zapatista." *alai*, July. https://ri.conicet.gov.ar/bitstream/ handle/11336/94112/CONICET_Digital_Nro.f1a1789b-ab5e-4c5d-8ee2- 60f37d0216b7_X.pdf?sequence=5.

575. Frison EA, et al. (2016). *From Uniformity to Diversity: A Paradigm Shift from Industrial Agriculture to Diversified Agroecological Systems*. International Panel of Experts on Sustainable Food Systems. http:// www.ipes-food.org/_img/upload/files/UniformityToDiversity_FULL. pdf; Anonymous (2014). *Regenerative Organic Agriculture and Climate Change*. Rodale Institute, Kutztown, PA http://rodaleinstitute.org/assets/ RegenOrgAgricultureAndClimateChange_20140418.pdf; Gliessman S, et al. (2018). *Breaking Away from Industrial Food and Farming Systems: Seven Case Studies of Agroecological Transition*. International Panel of Experts on Sustainable Food Systems. http://www.ipes-food.org/_img/upload/files/ CS2_web.pdf.

576. Wallace RG and R Kock (2012). "Whose food footprint? Capitalism, agriculture, and the environment." *Human Geography* 5(1): 63–83; Frison EA, et al. (2016). *From Uniformity to Diversity: A Paradigm Shift from Industrial Agriculture to Diversified Agroecological Systems*.

577. Chappell MJ (2018). *Beginning to End Hunger: Food and the Environment in Belo Horizonte, Brazil and Beyond*. University of California Press, Oakland.

578. Cox S (2020). *The Green New Deal and Beyond: Ending the Climate Emergency While We Still Can*. City Lights Books, San Francisco.

579. Malkan S (2018). "Pamela Ronald's ties to chemical industry front groups." U.S. Right to Know, 27 December. https://usrtk.org/our-investigations/pamela-ronald-led-chemical-industry-front-group-efforts; Malkan S (2019). "Cornell Alliance for Science is a PR campaign for the agrichemical industry." GRAIN, 27 November. https://www.grain.org/en/article/6368-cornell-alliance-for-science-is-a-pr-campaign-for-the-agrichemical-industry; Marris E (2020). "GMOs are an ally in a changing climate."

580. Marx K (1875). *Critique of the Gotha Program*. Progress Publishers, Moscow.

581. Caffentzis G (2013). *In Letters of Blood and Fire: Work, Machines, and the Crisis of Capitalism*. PM Press, Oakland; Wallace R (2018). *Canonical Instabilities of Autonomous Vehicle Systems*. Springer International Publishing, Cham, Switzerland.

582. U.S. Food Sovereignty Alliance (2000). "Home." U.S. Food Sovereignty Alliance, http://usfoodsovereigntyalliance.org/; Soul Fire Farm (2020). "Reparations." Soul Fire Farm, http://www.soulfirefarm.org/get-involved/reparations; Savanna Institute (2020). "About." Savanna Institute website, http://www.savannainstitute.org.

583. Cavooris R (2019). "Origins of the crisis: on the coup in Bolivia." *Viewpoint Magazine*, 18 November https://www.viewpointmag.com/2019/11/18/origins-of-the-crisis-on-the-coup-in-bolivia/.

To the Bat Cave

584. Zhong N, et al. (2003). "Epidemiology and cause of severe acute respiratory syndrome (SARS) in Guangdong, People's Republic of China, in February 2003." *The Lancet* 362(9393): 1353–1358; Xu R-H, et al. (2004). "Epidemiologic clues to SARS origin in China." *Emerg. Infect. Dis.*, 10(6): 1030–1037; deWit E, van Doremalen N, D Falzarano, and V Munster (2016). "SARS and MERS: recent insights into emerging coronaviruses." *Nat. Rev. Microbio.*, 14(8): 523–534.

585. Wang L and B Eaton B (2007). "Bats, civets, and the emergence of SARS." *Curr. Top. Microbiol. Immunol.* 315: 324–344.

586. Yuan J, et al. (2010). "Intraspecies diversity of SARS-like coronaviruses in *Rhinolophus sinicus* and its implications for the origins of SARS coronaviruses in humans." *J. Gen. Virol.*, 91(4). https://doi.org/10.1099/vir.0.016378-0.

587. Li W, et al. (2005). "Bats are natural reservoirs of SARS-like coronaviruses." *Science*. 310(5748): 676–679.

588. Drexler JF, VM Corman, and C Drosten (2014). "Ecology, evolution and

classification of bat coronaviruses in the aftermath of SARS." *Antiviral Res.* 101:45–56.

589. Corman VM, D Muth, D Niemeyer, and C Drosten C. (2018). "Hosts and sources of endemic human coronaviruses." *Adv Virus Res.* 100:163–188.

590. Mao X, G He, J Zhang, S Rossiter, and S Zhang (2013). "Lineage divergence and historical gene flow in the Chinese horseshoe bat (*Rhinolophus sinicus*)." *PLoS One* 8(2): E56786.

591. Ibid. Sun K (2019). "*Rhinolophus sinicus*." *The IUCN Red List of Threatened Species.* 2019: e.T41529A22005184. https://dx.doi.org/10.2305/IUCN. UK.2019-3.RLTS.T41529A22005184.en.

592. Mao X, G He, J Zhang, S Rossiter, and S Zhang (2013). "Lineage divergence and historical gene flow in the Chinese horseshoe bat (*Rhinolophus sinicus*)"; Liang L, X Luo, J Wang, T Huang, and E Li (2019). "Habitat selection and prediction of the spatial distribution of the Chinese horseshoe bat (*R. sinicus*) in the Wuling Mountains." *Environmental Monitoring and Assessment,* 191:4.

593. Feijó A, Y Wang, J Sun, F Li, Z Wen, D Ge, L Xia, and Q Yanga (2019). "Research trends on bats in China: A twenty-first century review." *Mamm Biol.* 98: 163–172.

594. Wu H, T Jiang, X Huang, and J Feng (2018). "Patterns of sexual size dimorphism of horseshoe bats: testing Rensch's rule and potential causes." *Scientific Reports* 8:2616. https://www.nature.com/articles/S41598-018-21077-7.

595. Kawamoto K (2003). "Endocrine control of reproductive activity in hibernating bats." *Zoological Science* 20(9):1057–1069.

596. Ye G-X, L-M Shi, K-P Sun, X Zhu, and J Feng (2009). "Coexistence mechanism of two sympatric horseshoe bats (*Rhinolophus sinicus* and *Rhinolophus affinis*) (Rhinolophidae) with similar morphology and echolocation calls." https://www.researchgate.net/publication/286370087_Coexistence_mechanism_of_two_sympatric_horseshoe_bats_Rhinolophus_sinicus_and_Rhinolophus_affinis_Rhinolophidae_with_similar_morphology_and_echolocation_calls.

597. Mao X, G He, J Zhang, SJ Rossiter, and S Zhang (2013). "Lineage divergence and historical gene flow in the Chinese Horseshoe Bat (*Rhinolophus sinicus*)." *PLoS ONE* 8(2): e56786; Mao X, G Tsagkogeorga VD Thong, and SJ Rossiter (2019). "Resolving evolutionary relationships among six closely related taxa of the horseshoe bats (*Rhinolophus*) with targeted resequencing data." *Molecular Phylogenetics and Evolution* 139:106551.

598. Magrone T, M Magrone, and E Jirillo (2020). "Focus on receptors for coronaviruses with special reference to angiotensin-converting enzyme 2 as a potential drug target—a perspective." *Endocr. Metab. Immune Disord. Drug Targets.* doi: 10.2174/1871530320666200427112902. Epub ahead of print.

599. Zheng Y-Y, Y-T Ma, J-Y Zang, and X Xie (2020). "COVID-19 and the cardiovascular system." *Nature Reviews Cardiology* 17:259–260; McGonagle D, JS O'Donnell, K Sharif, PE Emery, and C Bridgemwood (2020). "Immune

mechanisms of pulmonary intravascular coagulopathy in COVID-19 pneumonia." *Lancet Rheumatology.* Epub ahead of print. https://www.sciencedirect.com/science/article/pii/S2665991320301211.

600. Hou Y, Peng C, Yu M, Han Z, Li F, Wang L, Shi Z (2010). "Angiotensin-converting enzyme 2 (ACE2) proteins of different bat species confer variable susceptibility to SARS-CoV entry." *Arch. Virol.,* 155(10): 1563–1569.

601. Department of Communicable Disease Surveillance and Response (2003). Consensus Document on the Epidemiology of Severe Acute Respiratory Syndrome (SARS). World Health Organization, Geneva, Switzerland. https://www.who.int/csr/sars/en/WHOconsensus.pdf; Karlberg J, D Chong, and Y Lai (2004). "Do men have a higher case fatality rate of severe acute respiratory syndrome than women do?" *Am. J. Epidemiol.* 159(3): 229–231; Hamming I, et al. (2007). "The emerging role of ACE2 in physiology and disease." *J. Pathol.* 212(1): 1–11; Jin J-M, P Bai, W He, F Wu, X-F Liu, et al. (2020). "Gender differences in patients with COVID-19: Focus on severity and mortality." *Frontiers of Public Health,* 29 April. https://doi.org/10.3389/fpubh.2020.00152.

602. Yang J, Y Zheng, X Gou, K Pu, Z Chen, et al. (2020). "Prevalence of comorbidities and its effects in patients infected with SARS-CoV-2: a systematic review and meta-analysis." *International Journal of Infectious Diseases* 94:91–95; Baker MG, TK Peckham, and NS Seixas (2020). "Estimating the burden of United States workers exposed to infection or disease: A key factor in containing risk of COVID-19 infection." *PLoS ONE* 15(4): e0232452.

603. Rodriguez A (2020). "Texas' lieutenant governor suggests grandparents are willing to die for US economy." *USA Today,* 24 March. https://www.usatoday.com/story/news/nation/2020/03/24/covid-19-texas-official-suggests-elderly-willing-die-economy/2905990001/.

604. Wambier CG and A Goren (2020). "Severe acute respiratory syndrome coronavirus 2 (SARS-CoV-2) infection is likely to be androgen mediated." *J Am Acad Dermatol* 83:308–309; Goren A, S Vaño-Galván, CG Wambier, J McCoy, A Gomez-Zubiaur (2020). "A preliminary observation: Male pattern hair loss among hospitalized COVID-19 patients in Spain—A potential clue to the role of androgens in COVID-19 severity." *Journal of Cosmetic Dermatology,* 19:1545–1547; Wambier CG, S Vaño-Galván, J McCoy, A Gomez-Zubiaur, S Herrera, et al. (2020). "Androgenetic alopecia present in the majority of hospitalized COVID-19 patients—the 'Gabrin sign'." *J Am Acad Dermatol.* doi: 10.1016/j.jaad.2020.05.079. Epub ahead of print.

605. Hoffmann M, H Kleine-Weber, S Schroeder, N Krüger, T Herrler, et al. (2020). "SARS-CoV-2 cell entry depends on ACE2 and TMPRSS2 and is blocked by a clinically proven protease inhibitor." *Cell* 181(2):271–280.e8.

606. Ambrosino I, E Barbagelata, E Ortona, A Ruggieri, G Massiah, et al. (2020). "Gender differences in patients with COVID-19: A narrative review." *Monaldi Arch Chest Dis* 90(2):318–324.

607. Jordan-Young RM and K Karkazis (2019). *Testosterone: An Unauthorized Biography*. Harvard University Press, Cambridge, MA.

608. Gebhard C, V Regitz-Zagrosek, HK Neuhauser, R Morgan, and SL Klein (2020). "Impact of sex and gender on COVID-19 outcomes in Europe." *Biol Sex Differ.* 11: 29.

609. Ibid. Chen X, Ran L, Liu Q, Hu Q, Du X and X Tan (2020). "Hand hygiene, mask-wearing behaviors and its associated factors during the COVID-19 epidemic: A cross-sectional study among primary school students in Wuhan, China." *International Journal of Environmental Research and Public Health* 17(8): 2893; Gunasekaran GH, SS Gunasekaran, SS Gunasekaran, and FHBA Halim (2020). "Prevalence and acceptance of face mask practice among individuals visiting hospital during COVID-19 pandemic: Observational study." Preprint. doi:10.20944/preprints202005.0152.v1.

610. Shattuck-Heirdon H, MW Reiches, and SS Richardson (2020). "What's really behind the gender gap in Covid-19 deaths?" *New York Times,* 24 June. https://www.nytimes.com/2020/06/24/opinion/sex-differences-covid.html.

611. Brannstrom C (2009). "South America's neoliberal agricultural frontiers: Places of environmental sacrifice or conservation opportunity?" *Ambio* 38(3): 141–149.

612. Ambrosino I, E Barbagelata, E Ortona, A Ruggieri, G Massiah, et al. (2020). "Gender differences in patients with COVID-19: A narrative review."

613. Hou Y, C Peng, M Yu, Y Li, Z Han, et al. (2010). "Angiotensin-converting enzyme 2 (ACE2) proteins of different bat species confer variable susceptibility to SARS-CoV entry." *Archives of Virology,* 155:1563–1569.

614. Ibid. Ge X-Y, J-L Li, X-L Yang, AA Chmura, G Zhu, et al. (2013). "Isolation and characterization of a bat SARS-like coronavirus that uses the ACE2 receptor." *Nature* 503: 535–538.

615. Channappanavar R, C Fett, M Mack, PPT Eyck, DK Meyerholz, and S Perlman (2017). "Sex-based differences in susceptibility to Severe Acute Respiratory Syndrome coronavirus infection." *Journal of Immunology,* 198(10): 4046–4053.

616. Killerby M, H Biggs, C Midgley, S Gerber, and J Watson (2020). "Middle East Respiratory Syndrome coronavirus transmission." *Emerg. Infect. Dis.* 26(2). https://wwwnc.cdc.gov/eid/article/26/2/19-0697_article.

617. Hayman DTS, RA Bowen, PM Cryan, GF McCracken, TJ O'Shea, et al. (2013). "Ecology of zoonotic infectious diseases in bats: Current knowledge and future directions." *Zoonoses and Public Health,* 60(1): 2–21; O'Shea T, et al. (2014). "Bat flight and zoonotic viruses." *Emerg. Infect. Dis.* 20(5): 741–745; Subuhi S, N Rapin, and V Misra (2019). "Immune system modulation and viral persistence in bats: understanding viral spillover." *Viruses* 11(2): 192.

618. Fenton A, J Lello, MB Bonsall (2006). "Pathogen responses to host immunity: the impact of time delays and memory on the evolution of virulence." *Proc Biol Sci,* 273(1597): 2083-2090; Wyne JW and L-F Wang (2013). "Bats and viruses: Friend or foe?" *PLoS Pathogens* 9(10): e1003651.

619. Drexler JF, VM Corman, and C Drosten (2014). "Ecology, evolution and classification of bat coronaviruses in the aftermath of SARS"; Wong A, L Xin, S Lau, and P Woo (2019). "Global epidemiology of bat coronaviruses." *Viruses* 11(2): 174; Lau S, et al. (2019). "Novel bat alphacoronaviruses in Southern China support Chinese horseshoe bat as an important reservoir for potential novel coronaviruses." *Viruses* 11(5): 423.

620. Slingenbergh J and JM Leneman. Submitted. "Upon a virus host shift from wildlife to humans or livestock, the pathogenesis in the new host evolves to reflect the long term virus life history in the sylvatic cycle." *Viruses.*

621. O'Leary MA, JI Bloch, JJ Flynn, TJ Gaudin, A Giallombardo, et al. (2013). "The placental mammal ancestor and the post–K-Pg radiation of placentals." *Science* 339: 662–667.

622. Wallace RG. "Blood machines." This volume.

623. Smith KF, M Goldberg, S Rosenthal, L Carlson, J Chen, et al. (2014). "Global rise in human infectious disease outbreaks." *J R Soc Interface* 11(101): 20140950.

624. Field HE (2009). "Bats and emerging zoonoses: Henipaviruses and SARS." *Zoonoses and Public Health,* 56(6-7): 278–284; Wallace R, LF Chaves, LR Bergmann, C Ayres, L Hogerwerf, R Kock, and RG Wallace (2018). *Clear-Cutting Disease Control: Capital-Led Deforestation, Public Health Austerity, and Vector-Borne Infection.* Springer, Cham; Wallace RG, A Liebman, LF Chaves, and R Wallace (2020). "COVID-19 and circuits of capital." This volume; Fisher G (2020). "Deforestation and monoculture farming spread COVID-19 and other diseases." *Truthout,* May 12. https://truthout.org/articles/deforestation-and-monoculture-farming-spread-covid-19-and-other-diseases/.

625. Wallace RG, A Liebman, LF Chaves, and R Wallace (2020). "COVID-19 and circuits of capital."

626. Cahill P (2020). "U.S.-China blame game over COVID-19 heats up and Congress looks into antibody testing." NBC News, 29 April. https://www.nbcnews.com/news/morning-briefing/u-s-china-blame-game-over-covid-19-heats-congress-n1195146.

627. Arrighi G (2009). *Adam Smith in Beijing: Lineages of the 21st Century.* Verso, New York; Gulick J (2011). "*The Long Twentieth Century* and barriers to China's hegemonic accession." *American Sociological Association* 17(1): 4–38.

628. Shines R (2020) "WHO defunding threatens pillars of U.S. comprehensive national power." *Charged Affairs,* 25 May. https://chargedaffairs.org/who-defunding-threatens-pillars-of-u-s-comprehensive-national-power/.

629. Sylvers E and B Pancevski (2020). "Chinese doctors and supplies arrive in Italy." *Wall Street Journal,* 18 March. https://www.wsj.com/articles/chinese-doctors-and-supplies-arrive-in-italy-11584564673; Marques CF (2020). "China in Africa is more than a land grab." Bloomberg, 27 April. https://www.bloomberg.com/opinion/articles/2020-04-27/china-s-coronavirus-aid-to-africa-will-build-political-support.

630. Gretler C (2020). "Xi vows China will share vaccine and gives WHO full backing." Bloomberg, 18 May. https://www.bloomberg.com/news/articles/2020-05-18/china-s-virus-vaccine-will-be-global-public -good-xi-says.

631. Human Rights Watch (2020). "China: Covid-19 discrimination against Africans." 5 May. https://www.hrw.org/news/2020/05/05/china-covid-19-discrimination-against-africans; Abi-Habib M and K Bradsher (2020). "Poor countries borrowed billions from China. They can't pay it back." *New York Times*, 18 May. https://www.nytimes.com/2020/05/18/business/china-loans-coronavirus-belt-road.html; Marsh J (2020). "As China faces a backlash in the West, Xi needs Africa more than ever." CNN, 20 May. https://www.cnn.com/2020/05/19/asia/xi-jinping-africa-coronavirus-hnk-intl/index.html; George A (2020). "China's failed pandemic response in Africa." *Lawfare*, 24 May. https://www.lawfareblog.com/chinas-failed-pandemic-response-africa.

632. Anonymous (2020). "Africa's anti-COVID-19 efforts boosted by donations from China." Xinhua Net, 28 April. http://www.xinhuanet.com/english/2020-04/28/c_139015727.htm; Kapchanga M (2020). "African leadership shares China's health vision." *Global Times*, 20 May. https://www.globaltimes.cn/content/1188965.shtml; Kuyoh S (2020). "United fight against virus needed in Africa." *China Daily*, 25 May. http://global.chinadaily.com.cn/a/202005/25/WS5ecb248ea310a8b2411581e3.html.

633. Ward A (2020). "How China is ruthlessly exploiting the coronavirus pandemic it helped cause." *Vox*, 28 April. https://www.vox.com/2020/4/28/21234598/coronavirus-china-xi-jinping-foreign-policy.

634. People's Daily (2020). "US politicians reveal their cold-bloodedness in pandemic response." *Global Times*, 7 May. https://www.globaltimes.cn/content/1187657.shtml.

635. Valitutto MT, O Aung, KYN Tun, ME Vodzak, D Zimmerman, et al. (2020) "Detection of novel coronaviruses in bats in Myanmar." *PLoS ONE* 15(4): e0230802.

636. Silver A and D Cyranoski (2020). "China is tightening its grip on coronavirus research findings." *Nature* 580(7804): 439–440.

637. Alliance for Human Research Protection (2016) "Former FDA Commissioner is charged in RICO lawsuit." 23 April. https://ahrp.org/former-fda-commissioner-charged-in-federal-racketeering-lawsuit/.

638. Edelman M (2019). "Hollowed out Heartland, USA: How capital sacrificed communities and paved the way for authoritarian populism." *Journal of Rural Studies*, 10 November. https://www.sciencedirect.com/science/article/pii/S0743016719305157.

639. Wallace RG. "Midvinter-19." This volume.

640. Ye G-X, L-M Shi, K-P Sun, X Zhu, and J Feng (2009). "Coexistence mechanism of two sympatric horseshoe bats (*Rhinolophus sinicus* and *Rhinolophus affinis*) (Rhinolophidae) with similar morphology and echolocation calls."

641. Wu Z et al (2016). "Deciphering the bat virome catalog to better understand the ecological diversity of bat viruses and the bat origin of emerging

infectious diseases." *The ISME Journal* 10(3): 609–620; Fan Y, K Zhao, Z-L Shi, and P Zhou (2019). "Bat coronaviruses in China." *Viruses*, 11(3): 210; Zhang W (2018). "Global pesticide use: Profile, trend, cost/benefit and more." *Proceedings of the International Academy of Ecology and Environmental Sciences* 8(1): 1–27; Zhang W, Y Lu, W van der Werf, J Huang, F Wu, et al. (2018). "Multidecadal, county-level analysis of the effects of land use, Bt cotton, and weather on cotton pests in China." *PNAS* 115 (33): E7700-E7709; Maggi FF, HM Tang, D la Cecilia, and A McBratney (2019). "Global Pesticide Grids (PEST-CHEMGRIDS), 2015: Application rate of propanil on rice, high estimate." In *Global Pesticide Grids (PEST-CHEMGRIDS)*. Palisades, NY. https://sedac.ciesin.columbia.edu/downloads/maps/ferman-v1/ferman-v1-pest-chemgrids/ferman-v1-pest-chemgrids-app-rate-propanil-rice-high-est-2015.jpg.

642. Chen S and J Li (2020). "Xi says China won't return to planned economy, urges cooperation." Bloomberg, 23 May. https://www.bloomberg.com/news/articles/2020-05-23/xi-says-china-won-t-return-to-planned-economy-urges-cooperation.

643. Norton B and M Blumenthal (2019). "DSA/Jacobin/Haymarket-sponsored 'Socialism' conference features US gov-funded regime-change activists." *The Gray Zone*, 6 July. https://thegrayzone.com/2019/07/06/dsa-jacobin-iso-socialism-conference-us-funded-regime-change/.

644. La Botz D (2019). "Against the GrayZone slanders." *Medium*, 12 July. https://medium.com/@danlabotz/against-the-grayzone-slanders-ff305eecaf71.

645. Miranda A (2020). "COVID-19 Essentials pop-up offers in-demand supplies for pandemic in one place." WSVN News Miami, 11 June. https://wsvn.com/entertainment/covid-19-essentials-pop-up-offers-in-demand-supplies-for-pandemic-in-one-place/.

646. Feldman N (2020). "America has no plan for the worst-case scenario on Covid-19." Bloomberg, 6 May. https://finance.yahoo.com/news/america-no-plan-worst-case-153036385.html; Hawkins D, B Shammas, M Kornfield, M Berger, K Adam, et al. (2020). "Trump tells Oklahoma rally he directed officials to slow virus testing to find fewer cases." *Washington Post*, 20 June. https://www.washingtonpost.com/nation/2020/06/20/coronavirus-live-updates-us/; Fitz D (2020). "How Che Guevara taught Cuba to confront COVID-19." *Monthly Review*, 1 June. https://monthlyreview.org/2020/06/01/how-che-guevara-taught-cuba-to-confront-covid-19/; Kretzschmar ME, G Rozhnova, MCJ Bootsma, M van Boven, JHHM van de Wijgert, MJM Bonten (2020). "Impact of delays on effectiveness of contact tracing strategies for COVID-19: a modelling study." *The Lancet Public Health*, 16 July. https://doi.org/10.1016/S2468-2667(20)30157-2; Edwards E (2020). "Money and speed for COVID-19 tests needed to combat 'impending disaster.'" NBC News, 16 July. https://www.nbcnews.com/health/health-news/money-speed-covid-19-tests-needed-combat-impending-disaster-n1234037.

647. National Center for Health Statistics (2020). *Weekly Updates by Select Demographic and Geographic Characteristics: Provisional Death Counts for Coronavirus Disease 2019 (COVID-19)*. Centers for Disease Control and Prevention, 17 June. https://www.cdc.gov/nchs/nvss/vsrr/covid_weekly/index.htm#Race_Hispanic; Ford T, S Rener, and RV Reeves (2020). "Race gaps in COVID-19 deaths are even bigger than they appear." Brookings Institution, 16 June. https://www.brookings.edu/blog/up-front/2020/06/16/race-gaps-in-covid-19-deaths-are-even-bigger-than-they-appear/.

648. Ndii D (2020). "Thoughts of a pandemic, geoeconomics and Africa's urban sociology." *The Elephant*, 25 March. https://www.theelephant.info/op-eds/2020/03/25/thoughts-on-a-pandemic-geoeconomics-and-africas-urban-sociology/; Burki T (2020). "COVID-19 in Latin America." *The Lancet Infectious Diseases* 20(5):547–548.

649. Foster JB and I Sunwandi (2020). "COVID-19 and catastrophe capitalism: Commodity chains and ecological-epidemiological-economic crises." *Monthly Review* 72(2).

650. Whitehead MJ (2020). "Surveillance capitalism in the time of Covid-19: The possible costs of technological liberation from lockdown." Interdisciplinary Behavioral Insights Research Centre blog, 11 May. https://abi554974301.wordpress.com/2020/05/11/surveillance-capitalism-in-the-time-of-covid-19-the-possible-costs-of-technological-liberation-from-lockdown/; Timotijevic J (2020). "Society's 'new normal'? The role of discourse in surveillance and silencing of dissent during and post Covid-19." *Social Science & Humanities*, 27 May. https://papers.ssrn.com/sol3/papers.cfm?abstract_id=3608576; Hensley-Clancy M (2020). "The coronavirus is shattering a generation of kids." *Buzzfeed*, 11 June. https://www.buzzfeednews.com/article/mollyhensleyclancy/coronavirus-kids-school-inequality.

651. Taylor C (2020). "This map shows where coronavirus vaccines are being tested around the world." CNBC, 5 June. https://www.cnbc.com/2020/06/05/this-map-shows-where-coronavirus-vaccines-are-being-tested-worldwide.html.

652. Bienkov A (2020). "Scientists fear the hunt for a coronavirus vaccine will fail and we will all have to live with the 'constant threat' of COVID-19." *Business Insider*, 25 April. https://www.businessinsider.com/coronavirus-vaccine-may-be-impossible-to-produce-scientists-covid-2020-4; Collins F (2020). "Meet the researcher leading NIH's COVID-19 vaccine development efforts." *NIH Director's Blog*, 9 July. https://directorsblog.nih.gov/2020/07/09/meet-the-researcher-leading-nihs-covid-19-vaccine-development-efforts/; O'Donnell C (2020). "Merck CEO says raising COVID-19 vaccine hopes 'a grave disservice'–report." Reuters, 14 July. https://www.reuters.com/article/us-health-coronavirus-vaccine-merck-co/merck-ceo-says-raising-covid-19-vaccine-hopes-a-grave-disservice-report-idUSKCN24F2RV; Hollar J (2020). "Stories dooming vaccine hopes overlook immunity's complexity in search of easy clicks." FAIR, 22 July. https://fair.org/home/stories-dooming-vaccine-hopes-overlook-immunitys-complexity-in-search-of-easy-clicks/.

653. Lanese N (2020). "Researchers fast-track coronavirus vaccine by skipping key animal testing first." *Live Science*, 13 March. https://www.livescience.com/coronavirus-vaccine-trial-no-animal-testing.html; Megget K (2020). "What are the risks of fast-tracking a Covid-19 vaccine?" *Chemistry World*, 13 July. https://www.chemistryworld.com/news/what-are-the-risks-of-fast-tracking-a-covid-19-vaccine/4012130.article.

654. Peeples L (2020). "News Feature: Avoiding pitfalls in the pursuit of a COVID-19 vaccine." *PNAS*, 117(15): 8218–8221; Leming AB and V Raabe (2020). "Current studies of convalescent plasma therapy for COVID-19 may underestimate risk of antibody-dependent enhancement." *J Clin Virol.* 127: 104388; Garber K (2020). "Coronavirus vaccine developers wary of errant antibodies." *Nature Biotechnology*, 5 June. https://www.nature.com/articles/d41587-020-00016-w.

655. Wallace RG, R Kock, L Bergmann, M Gilbert, L Hogerwerf, C Pittiglio, R Mattioli, and R Wallace (2016). "Did neoliberalizing West Africa's forests produce a vaccine-resistant Ebola?" In Wallace R and RG Wallace (eds), *Neoliberal Ebola: Modeling Disease Emergence from Finance to Forest and Farm*. Springer International Publishing, Cham, pp 55–68.

656. Associated Press (2020). "Half of Americans would get a COVID-19 vaccine, AP-NORC poll finds." 27 May. https://www.nbcnews.com/health/health-news/half-americans-would-get-covid-19-vaccine-ap-norc-poll-n1215606.

657. Payne DC, SE Smith-Jeffcoat, G Nowak, U Chukwuma, JR Geibe, et al. (2020). "SARS-CoV-2 infections and serologic responses from a sample of U.S. Navy service members—USS Theodore Roosevelt, April 2020." *MMWR* 69(23): 714–721.

658. Long Q-X, X-J Tang, Q-L Shit, Q Li, H-J Deng, et al. "Clinical and immunological assessment of asymptomatic SARS-CoV-2 infections." *Nature Medicine,* 18 June. https://www.nature.com/articles/s41591-020-0965-6.

659. Grifoni A, D Weiskopf, SI Ramirez, J Mateus, JM Dan, et al. (2020). "Targets of T Cell responses to SARS-CoV-2 coronavirus in humans with COVID-19 disease and unexposed individuals." *Cell*, 181(7):1489–1501; Sekine, T, A Perez-Potti, O Rivera-Ballesteros, K Strålin, J-B Gorin, et al. (2020). "Robust T cell immunity in convalescent individuals with asymptomatic or mild COVID-19." bioRxiv, 29 June. https://www.biorxiv.org/content/10.1101/2020.06.29.174888v1.abstract.

660. Schultz PR and AI Meleis (1988). "Nursing epistemology: Traditions, insights, questions." *Journal of Nursing Scholarship* 20(4): 217–221; Reed PG (2006). "The practice turn in nursing epistemology." *Nursing Science Quarterly* 19(1):1–3.

661. Gulick J, J Araujo, C Roelofs, T Kerssen, M Figueroa, et al. (2020). "What is mutual aid? A COVID-19 primer." PReP Neighborhoods, Pandemic Research for the People, Dispatch 2, 14 May. https://drive.google.com/file/d/1f62eApKdHXCVa-rnHV-EjCPTEa4N6g9i/view.

662. Davis M (2018). *Old Gods, New Enigmas: Marx's Lost Theory*. Verso, New York.

663. Karatani K (2014). *The Structure of World History: From Modes of Production to Modes of Exchange*. Duke University Press, Durham, NC.

664. van der Sande M, P Teunis, and R Sabel (2008). "Professional and home-made face masks reduce exposure to respiratory infections among the general population." *PLoS ONE* 3(7): e2618; Tognotti E (2013). "Lessons from the history of quarantine, from plague to influenza A." *Emerg Infect Dis.*, 19(2): 254–259; Lynteris C (2018). "Plague masks: The visual emergence of anti-epidemic personal protection equipment." *Medical Anthropology* 37:6, 442–457; Bufano M and B Robbins (2020). "Facemasks through the ages, from medical aid to fashion statement." CBS News, 31 May. https://www.cbsnews.com/news/facemasks-through-the-ages-from-medical-aid-to-fashion-statement/; Zhang R, Y Li, AL Zhang, Y Wang, and MJ Molina (2020). "Identifying airborne transmission as the dominant route for the spread of COVID-19." *PNAS*. Advance publication, 11 June. https://doi.org/10.1073/pnas.2009637117.

665. Rieger MO (2020). "To wear or not to wear? Factors influencing wearing face masks in Germany during the COVID-19 pandemic." *Social Health and Behavior* 3(2):50–54.

666. Grandjean D, R Sarkis, J-P Tourtier, C Julien-Lecocq, A Benard, et al. (2020). "Detection dogs as a help in the detection of COVID-19: Can the dog alert on COVID-19 positive persons by sniffing axillary sweat samples? Proof-of-concept study." bioRxiv, 5 June. https://www.bioRxiv.org/content/10.1101/2020.06.03.132134v1.full.

667. Wallace RG, A Liebman, D Weisberger, T Jonas, L Bergmann, R Kock, and R Wallace. "The origins of industrial agricultural pathogens." This volume.

668. Wallace RG, A Liebman, LF Chaves, and R Wallace (2020). "COVID-19 and circuits of capital"; Tyberg J (2020). *Unlearning: From Degrowth to Decolonization*. Rosa Luxemburg Stiftung, May 2020. http://www.rosalux-nyc.org/unlearning-from-degrowth-to-decolonization/; Foster JB, T Riofrancos, L Steinfort, G Kallis, M Ajl, B Tokar, and H Moore (2020). "ROAR Roundtable: COVID-19 and the climate crisis." ROAR Magazine, 23 June. https://roar-mag.org/essays/roar-roundtable-covid-19-and-the-climate-crisis/.

669. Foster JB, T Riofrancos, L Steinfort, G Kallis, M Ajl, B Tokar, and H Moore (2020). "ROAR Roundtable: COVID-19 and the climate crisis." .

670. Wallace RG. "Midvinter-19." This volume; Chaw S-M, J-H Tai, S-L Chen, C-H Hsieh, S-Y Chang, et al. (2020). "The origin and underlying driving forces of the SARS-CoV-2 outbreak." *Journal of Biomedical Science* 27:73.

671. Brufsky A and MY Lotze (2020). "DC/L-SIGNs of hope in the COVID-19 pandemic." *Journal of Medical Virology*. https://doi.org/10.1002/jmv.25980.

672. Lee L, T Hughes, M-H Lee, H Field, Hume1, JJ Rovie-Ryan, et al. (2020). "No evidence of coronaviruses or other potentially zoonotic viruses in Sunda pangolins (*Manis javanica*) entering the wildlife trade via Malaysia." bioRxiv, 19 June. https://www.bioRxiv.org/content/bioRxiv/early/2020/06/19/2020.06.19.158717.full.pdf.

673. Yee E (2019). "The pangolin trade explained: situation in Malaysia." *The Pangolin Reports*, 2 May. https://www.pangolinreports.com/malaysia/.
674. Chan YA and SH Zhan (2020). "Single source of pangolin CoVs with a near identical Spike RBD to SARS-CoV-2." bioRxiv, 7 July. https://www.bioRxiv. org/content/10.1101/2020.07.07.184374v1; Thomas L (2020). "Research sheds doubt on the Pangolin link to SARS-CoV-2." *News Medical*, 8 July. https://www.news-medical.net/news/20200708/Research-sheds-doubt-on-the-Pangolin-link-to-SARS-CoV-2.aspx.
675. Latham J and A Wilson (2020). "A proposed origin for SARS-CoV-2 and the COVID-19 pandemic." *Independent Science News*, 15 July. https://www. independentsciencenews.org/commentaries/a-proposed-origin-for-sars-cov-2-and-the-covid-19-pandemic/.
676. Wallace RG. "The blind weaponmaker." Patreon. In preparation.
677. Wallace RG. "Midvinter-19." This volume.
678. Foster JB, T Riofrancos, L Steinfort, G Kallis, M Ajl, B Tokar, and H Moore (2020). "ROAR Roundtable: COVID-19 and the climate crisis."

Index